BIM and Quantity Surveying

The introduction of Building Information Modelling (BIM) as a key part of the building industry is redefining the roles and working practices of its stakeholders. Many clients, designers, contractors, quantity surveyors and building managers are still finding their feet in an industry where BIM compliance can bring great rewards.

This guide is designed to help quantity surveying practitioners and students understand what BIM means for them and how they should prepare to work successfully on BIM-compliant projects. The case studies show how firms at the forefront of this technology, have integrated core quantity surveying responsibilities like cost estimating, tendering and procurement into high-profile BIM projects. In addition to this, the implications for project management, facilities management, contract administration and dispute resolution are also explored through case studies, making this a highly valuable guide for those in a range of construction project management roles.

This book also considers how the role of the quantity surveyor is likely to significantly shift as a result of this development, as well as descriptions of tools used, this covers both the organisational and practical aspects of a crucial topic.

Steve Pittard FRICS is a lecturer in the School of the Built Environment and Architecture at London South Bank University where he is also an academic lead for the University's BIM Centre. As an active Fellow of the RICS, Steve is a member of the UK Technology Focus Group as well as an Assessor for the recently launched RICS BIM Manager Certification. Steve is also a co-founder of the BIM Academic Forum and an external examiner at the Dublin Institute of Technology.

Peter Sell MRICS is a partner at Gardiner & Theobald. He worked on the London 2012 and Crossrail programmes where he was involved in the development of the contractual BIM requirements. He has been involved in the development of technology use within quantity surveying for the last 25 years. Peter is an active member of the RICS: he sits on both the Infrastructure Forum and the UK Technology Focus Group and is an Assessor for the recently launched RICS BIM Manager Certification.

BIM and Quantity Surveying

Edited by
Steve Pittard and Peter Sell

Routledge
Taylor & Francis Group

LONDON AND NEW YORK

First published 2016
by Routledge
2 Park Square, Milton Park, Abingdon, Oxon OX14 4RN

and by Routledge
711 Third Avenue, New York, NY 10017

Routledge is an imprint of the Taylor & Francis Group, an informa business

British Library Cataloguing-in-Publication Data
A catalogue record for this book is available from the British Library

Library of Congress Cataloging-in-Publication Data
BIM and quantity surveying / Edited by Steve Pittard and Peter Sell.
pages cm
Includes bibliographical references and index.
ISBN 978-0-415-87042-9 (hardback : alk. paper) -- ISBN 978-0-415-87043-6
(pbk. : alk. paper) -- ISBN 978-1-315-67736-1 (ebook : alk. paper) 1. Building
information modeling. 2. Quantity surveying. I. Pittard, Steve, editor.
TH438.13.B55 2015
720.285--dc23
2015024378

ISBN: 978-0-415-87042-9 (hbk)
ISBN: 978-0-415-87043-6 (pbk)
ISBN: 978-1-315-67736-1 (ebk)

Typeset in Goudy Old Style
Servis Filmsetting Ltd, Stockport, Cheshire

Inspired by many learned colleagues, past and present, and by a shared passion and desire to demystify the mystified.

Also to our families for their support and understanding for the many hours of absence.

Contents

List of tables and figures

Tables

Figures

Foreword

The era of Building Information Modelling (BIM) has arrived. The question then for any Quantity Surveyor (QS) is will BIM elevate or demote the status of the QS?

To my mind this question is answered quite simply; that for those QSs that embrace BIM, their status will not only be secured for the future but also elevated. The QS using BIM will be relevant, in context with other members of the design team and offering an enhanced service to their clients. In my role as the future President of the Royal Institution of Chartered Surveyors, and as a fellow QS, I see many QSs, and, indeed, other construction professionals, grappling with the reality, practicality, and, indeed, opportunity that BIM will give them.

Technological change is especially important in construction. Although BIM is much more than technological tools, the reality for the QS is that they need to have the tools in their armour in order to remain competitive in a twenty-first-century marketplace. Embrace BIM or fall behind the curve of what the profession can provide and what your clients deserve.

What the authors have done here is provide a practical first step guide to BIM and how embracing what it can offer will help secure that future place for the QS. The book looks at the key service streams offered by the QS through a series of practical case studies. These case studies provide information on how BIM was, and can be, used, plus the issues and benefits of doing so.

The reality is this book is as relevant for other construction professionals wishing to better understand BIM as it is for the QS.

It is a fact that BIM offers great opportunity for the QS, but understanding 'how it will affect me' is the key first step to navigating a path that will lead to the benefits BIM will provide, most importantly, to our clients.

In *BIM and Quantity Surveying*, the lead editors, Steve Pittard (FRICS), lecturer at the School of the Built Environment and Architecture at London South Bank University where he is the academic lead for BIM; and Peter Sell (MRICS), a Partner with G&T, provide a great combination of academic and leading firm perspectives across this highly topical issue. Steve and Peter are senior well-regarded QSs having worked in the private sector, but also with a strong technological know-how within property and construction. They, together with some very strong case studies, bring the theory and practice of BIM into the realms of relevance for the QS of today. They are to be congratulated in doing so for the time for the QS to embrace the philosophy and technology of BIM is now, as the world of construction continues to move on at pace.

This is a well-thought-through book that will work for anyone seeking to get to grips with the practicalities of using BIM on their projects.

I was delighted to be asked to write the foreword for this book; as a QS myself I have always strived to be at the forefront of the profession and to embrace the latest technology on offer. This book is timely and it is relevant now. Once you have read this book I am sure you will be better equipped and enabled to embrace and understand the benefits (and some of the issues) that BIM can bring to the world of construction. I am passionate about our construction industry and embracing BIM as part of the next steps has to be the right thing for us as professionals, advisers and members of the built environment team.

Foreword by Amanda Clack MSC BSC FRICS
FAPM FIC FCMI CMC AFFILIATE ICAEW FRSA
Senior Vice President RICS

We live in a digital world where everyone and everything is connected.

For Quantity Surveyors, just as scale rules gave way to digitizers and they in turn to CAD measurement, the key tools for the next generation of our profession will be firmly rooted in the digital world of BIM.

Government is committed to the use of BIM on all of its projects and has spent time and resources making sure that they are ready for this across all departments (an area where the Construction Industry Council has played a major role). In the private sector, BIM is also now gaining momentum. Whether today we embrace it or not, BIM is with us and is here to stay.

Many QSs will say they are BIM ready or BIM compliant but I wonder how many of our profession can truly say that they are *fully* capable of operating in a BIM environment and really understand what it is like to provide professional services in a BIM world?

This book, whilst primarily aimed at the QS, is a guide that will assist all built environment professionals. Whether new to our industry or an experienced professional, this book is of interest to anyone wanting to know more about how BIM will affect them and is an important guide to industry practitioners.

Foreword by Tony Burton BSC FRICS MEWI
Chair of the Construction Industry Council

Preface

BIM and Quantity Surveying was the working title of the book from the very earliest days; as a title it simply stated what the book was to be about, how the implementation of BIM across the construction industry would affect the QS profession and the services that it delivers. As such the book takes a cross section of the discrete services that the QS offers to its clients and explores the effect of BIM on these services. Being straightforward professionals we saw no reason to move away from the simply stated title and purpose for the book.

As industry professionals of many years standing who have been involved in information technology since the 1980s and who were involved in the introduction of CAD and the introduction of electronic tools into the QS's office, we have seen a number of technology drives, all of which promised to deliver major benefits to the industry and the QS profession. Few, if any, achieved what they promised; all have nudged the 'supertanker' that is the industry a few degrees, but nothing like the revolution that they promised at the time. It is very clear that as an industry we are a 'late adopter' of technology, and on reflection, this is inevitable where margins are low and there is very little appetite for the risk that flows from the adoption of new technology and new ways of working.

BIM, however, is something different. BIM may be seen as geeky, but actually it is about so much more than some whizzy technology. Yes, it is technology that is enabling the change, and, yes, it is technology that is demanding that more thought be given to how information is structured. However, BIM is fundamentally about how we relate and collaborate as stakeholders on a project and how we work together to achieve the desired result for our clients. The mandating of Level 2 BIM for all UK Government projects by 2016 has given significant profile to BIM in the industry. This book attempts to reflect what has happened with regard to the services that the QS can and could deliver since the announcement of the target date by the UK Government in 2011.

Thus, the aim of this book is to inform, enlighten and educate the reader about what has happened with BIM in the industry and also to open their eyes to the possibilities that BIM brings to the QS as we move towards a BIM Level 3 world. We trust that this book will be as beneficial for the seasoned professional, as it will for the student embarking on their journey in what is one of the most important global industries.

The book has been written such that each chapter outlines a discrete service that either is, or could be, delivered by the QS. This means that each chapter stands alone. It would be helpful to read the introduction first to help set the scene for what follows, but after that the chapters can be read in the order which the reader finds most appropriate. The conclusion is relevant to any and all of the main chapters.

Acknowledgements

Permission to reproduce extracts from BSI publications is granted by BSI Standards Limited (BSI). No other use of this material is permitted. British Standards can be obtained in PDF or hard copy formats from the BSI online shop: <www.bsigroup.com/sShop>.

Abbreviations

AIM	Asset Information Model
AIR	Asset Information Requirements
APM	Association for Project Management
BEaaS	Built Environment as a Service
BEP	BIM Execution Plan (pre- and post-contract)
BIM	Building Information Modelling
CAD	Computer-Aided Design
CAFM	Computer-Aided Facilities Management
CDM	Construction Design and Management
CIC	Construction Industry Council
COBie	Construction Operations Building Information Exchange
DfMA	Design for Manufacture and Assembly
dPoW	Digital Plan of Work
EAM	Enterprise Asset Management System
ECC	The Engineering and Construction Contract
EVM	Earned Value Management
EIR	Employer's Information Requirements
FDC	Framework Design Consultants
FIDIC	International Federation of Consulting Engineers (a publisher of standard form contracts)
GSL	Government Soft Landings
IFMA	International Facilities Management Association
IP/IPR	Intellectual Property/Intellectual Property Rights
ITT	Invitation to tender
IWMS	Integrated workplace management system
JCT	Joint Contracts Tribunal (a publisher of standard form contracts)
KPI	Key performance indicators
LCC	Life cycle costing
LOD	Level of detail
MDM	Master Data Management
MIDP	Master Information Delivery Plan
MPDT	Model Production and Delivery Table

NEC	New Engineering Contract
OECD	Organisation for Economic Co-operation and Development
OIR	Organisational information requirements
OS	Ordinance Survey
PIM	Project information model/project Information Management
PLQ	Plain Language Questions
POE	Post-Occupancy Evaluation
PSC	Professional Services Contract
PTM	Project Team Member
RAMP	Risk Analysis and Management for Projects
QS	Quantity Surveyor
The CIC Protocol	The CIC's *Building Information Management Protocol*, first edition (2013), available free to download from www.cic.org.uk
RIBA	Royal Institution of British Architects
RIBA PoW	RIBA Plan of Work 2007
RICS	Royal Institution of Chartered Surveyors
SLA	Service Level Agreement

Contributors

Tahir Ahmad
Yemi Akinwonmi
Ian Aldous
Andrew R. Atkinson
Wes Beaumont
Phil Boyne
Julian Downes
Rob Garvey
Mark Kitching
Adrian Malone
Khalid Ramzan
Malcolm Taylor
Graeme White
Christopher Wright
Ahmed Zghari

Our gratitude and appreciation goes to all of those who devoted their expertise, experience and knowledge – not to mention of course their all-important time – to helping us write this book, without whom there would be no *BIM and Quantity Surveying*.

Our thanks also to Sue Pittard for providing the main chapter cartoons, to ensure we never take ourselves too seriously!

© SUE PITTARD

1 Introduction

Steve Pittard and Peter Sell

What this book is about

Whilst there is certainly no shortage of information about Building Information Modelling (BIM), there is little yet to guide industry practitioners and students through the maze of implementation and, more specifically, how BIM impacts their day-to-day roles and responsibilities – essentially putting BIM in the context of *their* world. The absence of this important missing link lies behind the core motivation of this text, addressing the all-important question of *how does BIM affect me*, and specifically *quantity surveying* (QS). Without this reference point, many industry professionals and students may struggle to make the required transition from general knowledge and awareness to full engagement and ability to deliver QS services in a BIM world.

Through a series of case studies, this book considers a number of key service streams to assess the impact of BIM on the role of the QS – essentially providing a practical guide to 'how does/will BIM affect me'. A list of recognised core QS service streams featured follows:

- Cost Planning
- Risk Management
- Whole Life Costing
- Procurement
- Information Management
- Contractual Framework for BIM
- Contract Administration
- Performance Measurement and Management
- Facilities Management
- Dispute Resolution.

Each service stream forms the basis of a separate chapter. However, it is also stressed that BIM is holistic and, therefore, the case studies are not intended to show fragmentation of process.

The book is not intended to either replace or duplicate existing texts on BIM or quantity surveying but, rather, to specifically help the reader gain an understanding of what BIM means for them, through specific service delivery examples, to provide relevance and context.

The service streams featured have been selected to reflect the broad nature and scope of a QS appointment. They illustrate the impact of BIM on some of the more traditional QS services such as Cost Planning, Measurement and Procurement, along with more contemporary and specialist services to highlight opportunities in the field of Performance and Information Management.

The case study format also lends itself more easily to revision as BIM matures and existing service streams change and/or new service streams come on line.

Who this book is aimed at

Although primarily aimed at Quantity Surveyors, the text is equally relevant for those carrying out any one of the services featured and, therefore, depending on the scope of service, should find interest with many others including Project Managers, Facilities Managers, Supply Chain Managers, Estimators, Commercial Managers and Building Surveyors. It is also intended to guide those just setting out on their career path as well as the more experienced professional. In essence, the book should find interest with anyone (practicing or otherwise) looking to gain a better understanding of how BIM does (or will) affect them.

Whilst the book features UK-based case studies, much of the content should also be relevant to those operating internationally as many of the principles should still apply. The book is also agnostic in terms of either project or business size and should, again, find equal relevance to both the large and the not so large!

Definitions

In order to ensure common meaning and understanding, the following establishes definitions of the two key terms used throughout this book, namely, *BIM* and *quantity surveying*.

The term facility and building are used ubiquitously to mean *any* built asset. The terms client and employer are interchangeable in this book.

What is BIM?

In its simplest form, the outcome of BIM is nothing more than delivering buildings or assets more efficiently – doing what we said we would do in the way we said we would do it and by the time we said we would deliver it! This outcome necessitates defining the format of, considering the usage of, and the processes associated with, the information required and produced from the construction of an asset (building) so that it is reusable for the entire life cycle of that asset. Whilst technology plays a role, BIM is more about the *way* we do things (process), embodying the ambitions of Latham and Egan (and other past evangelists for change) to drive improved efficiencies and eliminate waste.

BIM is in essence an enabler. By enabling better information flows, it enables better decision-making, which in turn enables better buildings and, consequently, better results.

In concept, BIM is extremely simple. The simplest definition, if rather sweeping 'single source of truth', appears on the Crossrail website (<http://www.crossrail.co.uk/benefits/design-innovation/>, accessed May 2015). However, even Gu and London's (2010) more complex definition quoted below reinforces the simplicity in concept:

> Building Information modelling (BIM) is an IT enabled approach that involves applying and maintaining an integral digital representation of all building information for different phases of the project lifecycle in the form of a data repository.
>
> (Gu and London, 2010: 988)

Another easily understandable definition, drawing on practice in the USA, comes from Azhar (2011):

> With BIM technology, an accurate virtual model of a building, known as a building information model, is digitally constructed.
>
> (Azhar, 2011: 241)

That said, both the definitions above imply a technology focus, which is misleading. After all, it is possible to have the best technology in the world and still be inefficient! The implementation of BIM requires a change in processes and relationships compared to non-BIM project delivery. BIM necessitates processes that support collaborative ways of working.

It is really all about changing the way we do things to improve performance and, more importantly, eliminate unnecessary (and avoidable) waste and inefficiency.

Essentially, BIM combines technology with new working practices to improve the quality of the delivered product and also improve the reliability, timeliness and consistency of the process to create, control and amend the information. It requires a move away from the traditional sequential workflow, to an environment where all parties share and effectively work with a common information pool – creating a 'single version of the truth'.

BIM changes the emphasis by making the model the primary repository for information, from which an increasing number of documents, or more accurately 'reports', such as plans, schedules, and bills of quantities may be derived. The primary asset of a Building Information Model is the information. In essence, BIM involves building a digital prototype of the asset or building and simulating it in a digital world – coining the phrase 'build before you build'.

BIM changes the traditional process by making the model the primary tool for the whole project team. This ensures that all the designers, contractors and

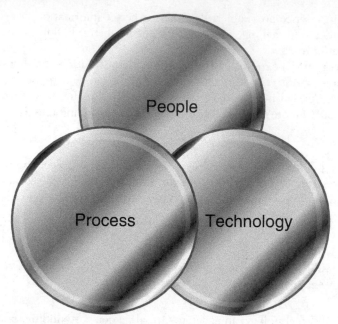

Figure 1.1 Key components of BIM.

subcontractors maintain a common basis for design and delivery so that the relationships between what have been traditionally isolated information pools may be explored and fully detailed. Working with BIM requires new skills and these will have to be learned from practice. As all parties involved with a BIM project have access to the same data, the information loss associated with handing a project over from design team to construction team and to building owner/operator is kept to a minimum – or even eliminated.

A building information model contains representations of the actual parts and components being used to construct a building along with geometry, spatial relationships, geographic information, quantities and properties of building components (for example manufacturers' details). BIM provides a common data environment (CDE) to store all the information that can be used to demonstrate the entire asset lifecycle from construction through to facility operation.

Often (mistakenly) referred to as 3D, 4D or nD, BIM should not be confused with the number of dimensions used to represent a building or asset. More traditional 2D or 3D drawings may well be outputs of BIM. However, instead of generating these in the conventional way (i.e. as individual drawings), they could all be produced directly from the model as a 'view' of the required information. In this regard, a better definition of BIM might actually be **Better Information Management!**

When defining BIM, it is often better to address the common misconceptions surrounding BIM by discounting some of the more popular – but inaccurate – definitions:

- It's **not** just 3D-CAD
- It's **not** just a new technology application
- It's **not** next generation, it's here and now!

(BIM Task Group, n.d.)

So, in many ways BIM is nothing new. People have conceptualised in 3D from the beginning of time and both architects and quantity surveyors are trained to think three dimensionally. Quantity surveyors are also trained to translate three-dimensional information into costs and construction managers to translate this same information into a programmed sequence. The difference with BIM is that much of this analytical process is removed to a single digital representation. This means both that it will no longer be necessary to imagine the information to be worked on as it will always be available in digital form, but also that many tasks (such as taking off quantities) will be automated or greatly deskilled.

BIM is not a panacea – it remains just as possible to produce a poor model, in terms of its functionality, its constructability or its value, as it is to produce poor drawings, schedules or any other, more traditional, form of information. Also, in the absence of any pro-active collaborative management effort, models may end up being prepared to suit the originator rather than being structured and presented with *all* parties to the design and construction team in mind. Ensuring that there is an agreed structure and exchange protocol in place to suit all parties will improve certainty, confidence and consistency. By moving to a shared information model environment, project failures and cost overruns become less likely. BIM certainly enables a better understanding and control of costs and schedules as well as providing an environment where the right information can be made available at the right time to reduce requests for information, manage change and limit (or even eliminate) unforeseen costs, delays and claims.

Clients are often in the best position to lead the introduction of BIM. Understanding the value of building information and its impact on their own business is leading many private sector clients to mandate BIM and to specify the standards and methods to be used in its adoption and this provides an opportunity for the QS in their capacity as one of the lead advisors where clients are yet to realise the potential of BIM for their business. The QS might also work with their clients to provide clear requirements for facilities management (FM) information to be handed over at project completion more easily with BIM. Some clients may begin to penalise lack of information (or the lack of its provision at established points in the construction process).

BIM is equally (and perhaps even more) applicable in the support of FM and asset management as it is to design and construction. Indeed, the output of the design model already has the capacity to replace the need for traditional operation and maintenance (O&M) manuals. Being able to interrogate an intelligent model, as opposed to searching through outdated paper-based O&M information, perhaps linked to interactive guidance on the repair and/or maintenance process, has obvious advantages. The potential for BIM in FM is discussed more fully in Chapter 10, where its use allows employers to take a longer-term view of the asset lifecycle.

However, the largest single barrier to exploiting BIM remains the lack of knowledge and awareness, and this again lies at the root and motivation for this book.

What is Quantity Surveying?

The definitions of Quantity Surveying can be many and varied.

According to the *Oxford English Dictionary*:

> a person who calculates the amount of materials needed for building work, and how much they will cost.
>
> oxforddictionaries.com

Whereas the RICS definition – taken from the 2014 pathway guide:

> Quantity surveyors are the cost managers of construction. They are initially involved with the capital expenditure phase of a building or facility, which is the feasibility, design and construction phases, but they can also be involved with the extension, refurbishment, maintenance and demolition of a facility. They must understand all aspects of construction over the whole life of a building or facility.
>
> As a quantity surveyor you may be working as a consultant in private practice, for a developer or in the development arm of a major organisation (eg. retailer, manufacturer, utility company or airport), for a public sector body or for a loss adjuster. On the contracting side you could be working for a major national or international contractor, a local or regional general contractor, for a specialist contractor or sub-contractor, or for a management style contractor.

[The] work [of a QS] may include the following:

- preparing feasibility studies or development appraisals
- assessing capital and revenue expenditure over the whole life of a facility
- advising clients on ways of procuring the project
- advising on the setting of budgets
- monitoring design development against planned expenditure
- conducting value management and engineering exercises
- managing and analysing risk
- managing the tendering process
- preparing contractual documentation
- controlling cost during the construction process
- managing the commercial success of a project for a contractor
- valuing construction work for interim payments, valuing change
- assessing or compiling claims for loss and expense and agreeing final accounts

- negotiating with interested parties
- giving advice on the avoidance and settlement of disputes.

(RICS, 2014a)

The RICS also define the QS as:

> an expert in the art of costing a building at all its stages.
> [...]
> Chartered Quantity Surveyors are highly trained professionals offering expert advice on construction costs. They are essential for life cycle costing, cost planning, procurement and tendering, contract administration and commercial management.
>
> (RICS, n.d.)

Yet another definition, offered by surveyors.com, defines quantity surveying as the following:

> Modern quantity surveyors provide services that cover all aspects of procurement, contractual and project cost management. They can either work as consultants or they can be employed by a contractor or sub-contract.
>
> (Surveyors.com, n.d.)

And the CIOB offers:

> Otherwise known as cost consultants or commercial managers, the role of a quantity surveyor is to keep a close eye on the various costs of a project, including materials, time taken and workers' salaries. They make sure that a construction project is as profitable and efficient as possible. They usually work for either the contractor (ie the company doing the building work), or the consultant or private quantity surveyor (ie the firm employed to advise the client).
>
> (http://www.ciob.org accessed April 2013)

It is perhaps interesting to note the absence of any direct reference to the term 'Quantity Surveyor' on the websites and marketing collateral of many larger UK practices, preferring to position themselves through the range of functions and portfolio of services provided.

Whilst this book features case studies drawn from the UK, much of the content can be applied internationally as many of the principles are common to the function and role of the QS or cost consultant worldwide.

Background

So how did BIM come about? Is BIM really all that new, and why do we need it? Well, it might be worth recapping some of the history of BIM to answer these questions to provide some background and context.

Whilst many other industries have seen considerable change, construction has seen comparably little, continuing to operate processes and procedures which have remained largely unchanged since the days of the Master Builder – accepting a level of productivity which, to put it mildly, can at best be described as poor. Industries such as Retail and Manufacturing have embraced change and the digital age, replacing outdated practices and revolutionising the way they do business, bringing significant efficiency and productivity improvements.

One clear example of this can be seen from the introduction of barcoding in the retail sector. Introduced in the 1970s, barcoding became a game changer in the way the retail sector operated. Initially used to capture item cost, its use was quickly extended as the potential to track products through the supply chain and on to the point of sale became obvious.

Despite the fact that digital technology has the potential to revolutionise construction and the long-term management of the built environment, repeated calls to embrace change have, thus far, been pretty much ignored.

Whilst BIM may be a relatively new term, the principle concepts of collaborative working and new ways of working are evident in reports such as the Placing and Management of Contracts for Building and Civil Engineering Work (otherwise known as the Banwell Report) (Banwell, 1964), *Constructing the Team* (Latham, 1994), *Rethinking Construction* (Egan, 1998) and *Accelerating Change* (Egan, 2002). Whilst these reports were endorsed by industry, it did little to embrace the changes recommended, essentially perpetuating the inefficiencies resulting from the embedded attributes of fragmentation and silo working.

So, why now? Well, in truth, this is probably largely down to one of the worst economic crises in recent history. Her Majesty's Government (HMG) saw BIM as a means to an end – the means to significantly reduce the capital cost of delivering new facilities – or to put it another away, delivering more for less. Whether it be delivering on their Building Schools for the Future (BSF) programme, or Procure21, BIM offered a real opportunity to get more for their (and our) buck! An attractive proposition to say the very least, for, what was at the time (2010), a cash-strapped administration.

Having been convinced of the potential, the Government incorporated BIM into its Construction Strategy (2011), and established the BIM Task Group (www.bimtaskgroup.org) to support the delivery of BIM. The website summarises its strategy for BIM:

> The Government's four-year strategy for BIM will change the dynamics and behaviours of the construction supply chain, unlocking new, more efficient and collaborative ways of working. This whole sector adoption of BIM will put us at the vanguard of a new digital construction era and position the UK to become the world leaders in BIM.
>
> (Francis Maude, Minister for the Cabinet Office)

HMG were confirming what the industry already knew, that the construction industry – at least in a process and systems context, was broken, and moreover, that a new way of working was required if the industry was going to meet the demands of society in the twenty-first century.

So BIM was formally launched in the UK following the UK Government's Construction Strategy in 2011 stating that BIM would be mandated on UK centrally procured public sector projects initially with a capital value in excess of £5m by 2016 (Cabinet Office, 2011). This limit was subsequently removed to apply to *all* UK centrally procured public sector projects and institutions and standard form publishers have since been busy developing standards, protocols and contractual frameworks to enable BIM (that is BIM 'Level 2') deployment on projects in the UK construction industry.

So does BIM only affect projects in the public sector? Many private sector clients have seen the potential of BIM and, in some cases, are ahead of the UK Government in terms of mandating BIM. Retail clients such as John Lewis and Asda were among the first in the private sector to mandate BIM – and some two years ahead of the Government target.

Other sectors have also been actively embracing the concepts of BIM, instigating a number of BIM4 forums such as BIM4infrastructure, BIM4SMEs and BIM4Rail (www.bimtaskgroup.org).

The Netherlands, Denmark, Finland and Norway also mandate BIM for public sector procurement with other Euro member states looking to follow suit. The growing use of BIM internationally opens up new export opportunities for UK design and construction businesses able to offer the required experience and expertise.

However, while the concepts and philosophy of BIM may be very simple, in its implementation, BIM has become perhaps unnecessarily complex. It has fostered its own language, jargon and plethora of three-letter acronyms, which to the uninitiated can seem confusing and may even present a barrier to wider adoption. Some software vendors may also be guilty of perpetuating this complexity through their failure to fully embrace open standards and by fuelling the market (and need) for the 'BIM specialist'.

So let's address this complexity by explaining some of the basics and key terms to help you understand BIM.

BIM jargon explained

Under the BIM Task Group, HMG adopted a 'push/pull' strategy to support the approach under BIM, providing the required demand (pull) through key delivery targets, which included a 20 per cent (real terms) reduction in capital delivery costs. However, they stopped short of tampering directly with the 'supply chain' engine, committing instead to providing the necessary standards and protocols to support the required industry 'push' to meet the demands of a BIM approach.

This has led to a number of new key standards and protocols, providing the necessary 'tools' to support BIM delivery:

- A new Digital Plan of Work (dPoW) (replacing the RIBA Plan of Work along with its many other sector-specific equivalents)
- BS1192:2007 (Information Production)
- PAS1192-2:2013 (Capital Delivery)
- PAS1192-3:2014 (Operational Delivery)
- BS1192 Part 4:2014 (COBie)
- PAS1192-5:2015 (Security)
- CIC BIM Protocol
- (Government) Soft Landings (BSRIA, 2009)
- Classification – Uniclass 2, etc. (in progress at the time of publication)

At the time writing, the UK Government were in the process of procuring a Digital Tool for BIM (NBS BIM Toolkit). It was understood that this new tool will incorporate the Digital Plan of Work mapped to the information delivery cycle (see Figure 1.3) and be freely available to guide the delivery of BIM. This reinforces the Government's commitment to BIM and its policy of removing

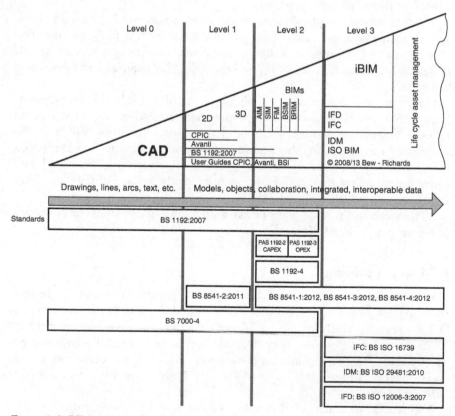

Figure 1.2 BIM maturity levels.
Bew and Richards (2008)

possible barriers, which might otherwise restrict industry take-up through the provision of freely available open standards and supporting tools for adopting BIM.

The implementation of BIM for projects has been conceptualised as a series of maturity levels (see Figure 1.2 for BIM maturity model) under the Construction Strategy published in 2011 by the UK government. BIM Level 2 was mandated by the 2011 Construction Strategy on all centrally procured UK Government construction projects by 2016.

Level 2 was subsequently defined in 2014 as a set of requirements. These requirements include adoption of a process set out by two BSI published documents: PAS1192-2:2013 and PAS1192-3:2014. A series of supplementary or supporting documents to the PAS1192 document set are also included in the Level 2 definition.

The information delivery life cycle (Figure 1.3) underlines the importance of a common and consistent structured process, ensuring correct and timely engagement of all parties through both the capital and operational life cycle of an asset.

The information delivery life cycle commences with a clear articulation of the project business case. By identifying the business case, a strategic plan for delivery may be created, which is not dissimilar to the traditional process. In the context of BIM, this is referred to as the Employer's Information Requirements (EIRs) (see also Chapter 5).

The EIR sets out the organisational information requirements necessary to manage the asset design, construction and operation lifecycle. The EIR will also contain a number of Plain Language Questions (PLQs), which offer an

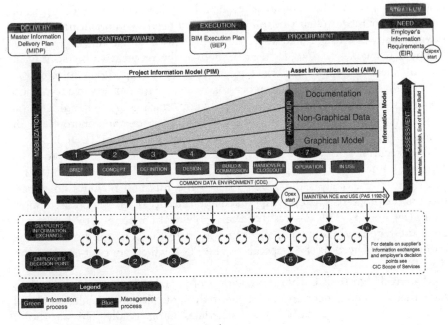

Figure 1.3 The information delivery cycle.

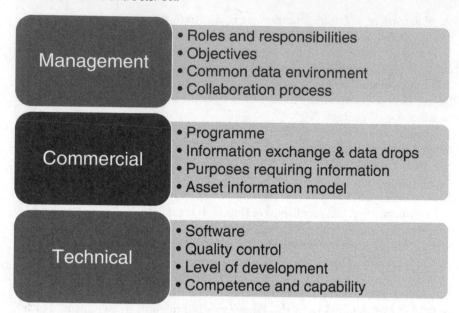

Figure 1.4 Employer's Information Requirements.

opportunity to guide the project team in terms of their information delivery. Presenting a question, theoretical or otherwise, at specific stages of the project lifecycle, enables all stakeholders to understand the process of that stage. For example, at a capital expenditure gateway, a PLQ may refer to the cost plan or target cost, thereby explicitly defining a particular need for information avoiding the ambiguity often associated with a non-BIM delivery environment.

The EIR enables the project team to deliver the project, whilst ensuring it not only meets the employer's brief but also delivers the appropriate information in line with the business case. The process is evolving with industry in a transition from static information in the form of hard copy drawings and schedules to the concept of the provision of information, which is dynamic and appropriate for each step in the delivery process, referred to as the level of detail (LOD).

The execution of BIM during a project, including the procedures and processes adopted, is referenced in the pre- and post-contract BIM Execution Plan (BEP). The BEP is effectively a direct response to the EIR and includes an assessment of the proposed capability, approach and capacity of the responding supply chain (Figure 1.5). This ensures that the supply chain appointed to deliver can meet the requirements of the project and assist in the accomplishment of the business case. The BEP also responds to the PLQs which are contained within the EIR.

Modern design tools enable the creation of parametric information models, such that an amendment to one element dynamically amends all associated elements, drawings, specifications and schedules automatically. This also holds true in respect of contractual documentation and removes the requirement for

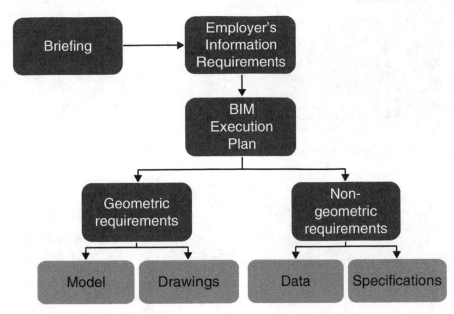

Figure 1.5 BIM execution.

multiple information delivery plans. For example, having a master programme, a procurement schedule, an information required schedule, a design responsibility matrix, etc., is unnecessary if the information is held within the model with the appropriate links between the information enabling dynamic coordination. Having a model progression delivery table (MPDT), akin to Figure 1.8, enables the creation of a single document, which sets out the information requirements for the entire delivery process.

As a project progresses through its life cycle, the information made available increases (Figure 1.6). By creating and sharing this information in a structured and consistent format for the use of all stakeholders, the value of this shared knowledge base is maximised to support improved decision making throughout the information delivery life cycle. At its most fundamental level, the benefit of BIM lies in the ability to reuse information for a number of purposes, identified by the business case and supplemented by uses throughout the project life cycle.

Information requirements are governed by the LOD, which have specific purposes at each maturity level, illustrated in Figure 1.7. By clearly and explicitly identifying and articulating the information requirements at each stage, using the LODs, it is possible for all stakeholders to be fully aware of their roles, responsibilities and obligations at key milestones.

Figure 1.7 provides an example of a specific purposed LOD classification, which is intended to provide greater clarity on information use. This structure is then emulated through the entire project and across all stakeholders. By identifying information needs and uses for all stakeholders, this helps ensure that all

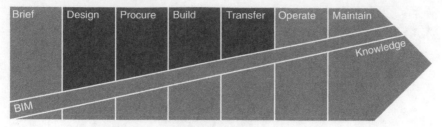

Figure 1.6 Increased information and knowledge.

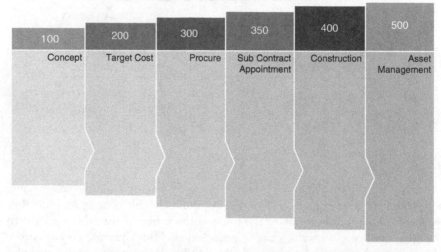

Figure 1.7 Level of detail and project uses.

RIBA 2013 Workstage	Preparation & brief 1		Concept 2		Developed 3		Technical Design 4	
Elements, Materials & Components (NRM1)	LOD	Resp Party	LOD	Resp Party	LOD	Resp Party	LOD	Resp Party
1 Substructure								
1 Substructure 1 Standard Foundations			200	CSE	300	CSE	300	CSE
2 Specialist Foundations			200	CSE	300	CSE	300	CSE
3 Lowest Floor Construction			200	CSE	300	CSE	300	CSE
4 Basement Excavation			200	CSE	300	CSE	300	CSE
5 Basement walls and retaining walls			200	CSE	300	CSE	300	CSE
2 Superstructure								
1 Frame 1 Steel frames			200	CSE	300	CSE	300	CSE
2 Space frames/decks								
3 Concrete casings to steel frames			200	CSE	300	CSE	300	CSE
4 Concrete frames								
5 Timber frames								
6 Specialist frames								
2 Upperfloors 1 Floors			200	CSE	300	CSE	300	CSE
3 Roof 1 Roof structure			200	CSE	300	CSE	300	CSE
2 Roof coverings			200	ARC	300	ARC	350	ARC
3 Specialist roof systems			200	ARC	300	ARC	350	ARC
4 Roof drainage			200	ARC	300	ARC	350	ARC
5 Rooflights, skylights and openings			200	ARC	300	ARC	350	ARC
6 Roof features			200	ARC	300	ARC	350	ARC
4 Stairs and ramps 1 Stair/ramp structures			200	CSE	300	CSE	300	CSE

Figure 1.8 Model progression delivery table – overview (extract).

information created fit for purpose, speeding up decision making and reducing – or even eliminating – uncertainty and ambiguity.

During design, the regular sharing of project information in a CDE enables the project team to work concurrently, which can reduce the design period. Figure 1.8

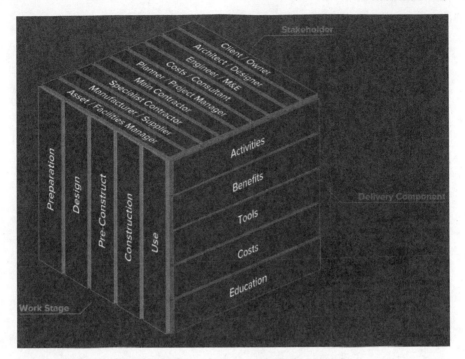

Figure 1.9 The BIM Delivery Cube.
Reproduced with permission of CIRIA and BIM TA.

shows a model progression delivery table which can be used as part of this process. The use of a common information exchange protocol also means that outputs from one member of the project team can be utilised as inputs for other members of the project team, be that cost managers, project managers or contract administrators, avoiding unnecessary duplication, reformatting and/or rework often associated with traditional workflow.

In order to ensure the efficient use and collation of project information, the BIM Task Group offers the concept of a 'data cube', used to access and display three axes representing work stages, stakeholders and data to illustrate the principles of a unified classification system. The approach is shown in Figure 1.9.

By 'slicing' the cube along any combination of axes, it is possible to reveal information that is relevant to each role, stage or data type, removing all other irrelevant information, enabling the presentation of a simple concise view of the project in specific user terms (CIRIA, 2014).

Case study approach

In an attempt to provide consistency and relevance, case studies used throughout this book generally follow a similar outline structure. Whilst some are based on real experience, others, where BIM maturity is still limited, are based on the views of experts in their field of how BIM is likely to impact those carrying out similar services.

A list of core QS services featured is provided earlier in this chapter.

Outline case study structure

Where chapters are based on real experience, content is generally framed around the following outline structure:

- Introduction:
  - Brief outline of content
  - Author bio
  - Company/project information
- Service profile:
  - Description of the core service provided, including any specific departures and/or specialisms relative to the case study
  - Scope of service
  - Whether BIM impacted the service, and if so, how?
  - Definitions
- Case study details:
  - Names of parties where available and/or relevant
  - Details of project or project(s) if describing generic services offered
  - Project value(s)
  - Dates
  - Why BIM was used and who made the decision
- How BIM was used:
  - Description of how BIM was used
  - Impact on traditional process(es)
- Tools used:
  - A description of the tool(s) used – generic rather than proprietary
  - How tool(s) were selected/implemented
- Issues/benefits:
  - A summary of any issues or problems which had to be overcome
  - How any issues/problems were addressed
  - Standards – are there any that are appropriate?
  - Standards – were they any good?
  - A summary of the key benefits resulting from the use of BIM:
    - To the project
    - To the business
    - To others
  - Were these benefits measured, and if so, how?
  - How might these benefits be repeated?
  - Lessons learned
- Summary/commentary:
  - Brief summary of the case study

- Any further commentary on either the case study or benefits/issues, etc.
- References

Where there was no experience of BIM, chapters have been framed around the following outline structure:

- Introduction:
  - Brief outline of content
  - Author bio
  - Company/project information
- Service profile traditional:
  - Description of the core service provided, including any specific departures and/or specialisms relative to the case study
  - Traditional scope of service
  - Definitions
- How BIM could be used:
  - Description of how BIM could be used
  - Impact on traditional process(es)
  - Additional or adjustments to scope of service
- Tools capability:
  - A description of the tool(s) that could be used – generic rather than proprietary
  - How tool(s) were selected/implemented
- Issues / benefits:
  - A summary of any issues or problems which need to be overcome
  - How any issues/problems could be addressed
  - A summary of the key benefits that could result from the use of BIM:
    - To projects
    - To the businesses
    - To others
  - Could these benefits measured, and if so, how?
  - How might these benefits be repeated?
  - Lessons to be learned
- Summary/commentary:
  - Brief summary of the case study
  - Any further commentary on either the case study or benefits/issues, etc.
- References

There are three chapters where this structure has been modified, *Risk and Risk Management*, *Contractual Framework for BIM* and *Information Management*, either due to the more general nature of the subject area (in the case of the first two) or because the role has yet to mature into a recognised service stream (in the case of the latter).

References

Azhar, S. (2011) Building Information Modeling (BIM): trends, benefits, risks, and challenges for the AEC industry, *Leadership and Management in Engineering*, 241–252, July.

Banwell, H. (1964) *Report of the committee on the placing and management of contracts for building and civil engineering work*, HMSO, London.

Bew, M. and Richards, M. (2008) BIM Maturity Model – BSI PAS1192-2:2013.

BIM Task Group (n.d.) 'FAQs. Bim Task Group'. Online. Available HTTP: <http://www.bimtaskgroup.org/bim-faqs/>, accessed 9 November 2014.

BSI (2007) *BS1192 Part 1 Collaborative production of architectural, engineering and construction information code of practice*, British Standards Institute, London.

BSI (2013) *PAS1192-2 Specification for information management for the capital/delivery phase of construction projects using building information modelling*, British Standards Institute, London.

BSI (2014) *PAS1192-3 Specification for information management for the operational phase of assets using building information modelling*, British Standards Institute, London.

BSI (2014) *BS1192 Part 4 Collaborative production of information: fulfilling employer's information exchange requirements using COBie – code of practice*, British Standards Institute, London.

BSI (2014) *PAS1192-5 Specification for security-minded building information management, digital built environments and smart asset management*, British Standards Institute, London.

BSRIA BG 4 (2009) *Report BG 4/2009 The soft landings framework*, Building Services Research and Information Association, Bracknell, UK.

Cabinet Office (2011) *Government construction strategy*, HMSO, London. Online. Available HTTP: <http://www.cabinetoffice.gov.uk/resource-library/government-construction-strategy>, accessed May 2015.

CIC (2013) *BIM Protocol*, Construction Industry Council, London.

CIRIA (2014) 'The BIM Delivery Cube'. CIRIA. Online. Available HTTP: <http://www.ciria.com/bimcube.html>, accessed May 2015.

Crossrail (2014) 'Driving industry standards for design innovation on major infrastructure projects'. Crossrail. Online. Available HTTP: <http://www.crossrail.co.uk/benefits/design-innovation/>, accessed May 2015.

Egan, J. (1998) *Rethinking Construction: report from the Construction Task Force*, Department of the Environment, Transport and Regions, London.

Egan, J. (2002) Accelerating Change: a report by the Strategic Forum for Construction Chaired by Sir John Egan, Rethinking Construction, London

HM Government (2013) *BIS/13/955 Construction 2025 industrial strategy: government and industry in partnership*, HM Government, London. Online. Available HTTP: <https://www.gov.uk/government/publications/construction-2025-strategy>, accessed May 2015.

Gu, N. and London, K. (2010) Understanding and facilitating BIM adoption in the AEC industry, *Automation in Construction* 19, 988–999.

Latham, M. (1994) *Constructing the team: final report of the government/industry review of procurement and contractual arrangements in the UK construction industry*, HMSO, London.

NBS (n.d.) 'BIM toolkit'. NBS. Online. Available HTTP: <www.thenbs.com/bimtoolkit/>, accessed May 2015.

RICS (2014a) 'Construction pathway guide'. RICS. Online. Available HTTP: <http://www.rics.org/uk/apc/pathway-guides/construction-pathway-guides/quantity-surveying-and-construction/>, accessed May, 2015.

RICS (2014b) *New Rules of Measurement 3: Order of Cost Estimating and Cost Planning for Building Maintenance Works*, First Edition, RICS, London

RICS (n.d.) 'Glossary'. RICS. Online. Available HTTP: <http://www.rics.org/uk/footer/glossary/quantity-surveying/>, accessed November 2014

Surveyors.com (n.d.) 'Definition of quantity surveyors'. Surveyors.com. Online. Available HTTP: <http://www.surveyors.com/quantity-surveyors/definition-of-quantity-surveyors-/>, accessed November 2014.

© SUE PITTARD

2 Cost planning

Ahmed Zghari and Ian Aldous

Introduction

This chapter combines a number of individual case studies to illustrate how the use of BIM can impact the preparation and delivery of a cost plan.

Whilst the purpose of this chapter is not to provide a reworked summary of established cost planning techniques and guidance within the *RICS New Rules of Measurement, Volume 1* (NRM1): *Order of Cost Estimating and Cost Planning for Capital Building Works*, it is perhaps important to establish the broad scope and purpose of service often provided. With the introduction of the *RICS New Rules of Measurement, Volume 3* (NRM3): *Order of Cost Estimating and Cost Planning for Building Maintenance Works* and a greater focus on costs for lifetime ownership, cost planning can be considered as a financial planning tool to provide the employer with an indication of the most likely cost of completing a project and the lifetime operational costs. A breakdown of the elemental components of the expected costs within a plan can then be used by employers to make informed decisions.

Cost planning is used by employers to set cost parameters, which are typically influenced by market-driven budgetary constraints or funding awards, both of which are derived from a review of financial viability. These then provide a framework around which a series of decisions can be made by the project teams, such as:

- Design and value considerations
- Identifying and mitigating risk factors
- Procurement options
- Project planning and programming
- Financial planning, cash flow forecasting and funding
- Creating project plans for cost monitoring, cost auditing and cost control.

The purpose and approach to cost planning is a subject that has been very well defined. The RICS NRM1 provides a detailed review of cost planning, the measurement rules and other considerations such as grants and capital allowances which may need to be identified while providing a cost planning service.

As well as illustrating how BIM can impact cost planning, this chapter also offers a practical review of:

- How BIM can be used to enhance best practice cost planning techniques;
- The benefits and challenges that can arise during the process; and
- How BIM is likely to change the practical approach to cost planning and collaborative engagement with the design team.

As the technology available for use in cost planning improves, the practical experiences gained by the authors during the early estimating and outline cost planning stages has highlighted the potential for further improvements in productivity with an increasing shift away from the technical challenges of generating quantities to a focus on cost planning, cost control, risk management and procurement planning.

The BIM case studies outlined in this chapter show how cost planning services might evolve. This evolution has been accelerated by the UK Government BIM mandate, which in itself has had a significant impact given that the public sector accounts for around half of the UK construction industry expenditure.

Author biographies

Ahmed Zghari

Ahmed Zghari trained as a contracts quantity surveyor with Alfred McAlpine and then within the real estate division of accountancy and consulting practice Arthur Andersen. After transferring to Savills Commercial as Associate Director to work within the cost consulting unit, he worked in the finance and retailing sector. He joined HCQS shortly after the government announced it would mandate BIM as part of its Construction Strategy.

Ahmed gained an honours degree in quantity surveying from the University of the West of England in 1996 and became a member of the RICS in 1998.

Ian Aldous

Ian Aldous is an Associate at Arcadis and a Chartered Surveyor, as well as a RICS Certified BIM Manager. Ian has significant experience delivering projects as both a Cost and Project Manager; he also provides strategic advice to clients on BIM implementation.

Ian has worked with a variety of clients to develop BIM initiatives, including BAA at Heathrow and the Welsh Government on a highways scheme. He has also delivered BIM services on complex, highly engineered facilities such as the National Graphene Institute for the University of Manchester.

Ian has been part of various BIM working groups including the RICS group and the CIC's BIM2050 team. He has also worked to help develop the UK Government standards on commercial elements of BIM.

Company information

HC QS

Originally established in 1989 as Haleys, HC QS operates in the UK and internationally providing quantity surveying, M&E design, project management, cost modelling and legal support services. In 2009, HC QS established BIM cost consulting as a specialist service within the group.

Particular areas of sector focus include Infrastructure; Building services; Healthcare; and Housing.

Arcadis

Arcadis is a design and consultancy for natural and built assets. They have been using BIM since its infancy for a range of deliverables on projects, including strategic optioneering and appraisal analysis; estimating and bill of quantities; elemental cost plans; procurement and tendering; cost management; project management; change management; programme management; and asset performance and optimisation.

Case studies

The case studies featured in this chapter are:

- **New capital building works.** Foundations and enabling works, King's College Hospital NHS Trust, London (Case Study 1: HC QS).
- **Works within existing building.** Remodelling and refurbishment feasibility project at Waterloo Station. Network Rail, London Waterloo [Case Study 2: HC QS].
- **New infrastructure works.** Major civil engineering earthworks. European Rail Operator, Location not specified (Case Study 3: HC QS).
- **Research facility.** New-build, highly engineered research facility for the University of Manchester (Case Study 4: Arcadis).

The changing nature of technology and the increased levels of expenditure in new tools since the government announced its BIM strategy in November 2011 are likely to have a continuing impact on all aspects of the delivery process. These case studies are, therefore, not only intended to show how BIM tools can be used but also to be indicative of the changing nature of the approach to cost planning.

Case Study 1: King's College Hospital NHS Trust, Denmark Hill, London

Service profile

The service provided was traditional quantity surveying advice to the client and the design team.

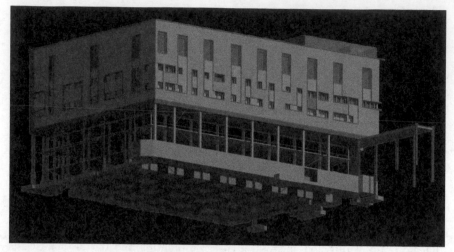

Figure 2.1 3D model view of King's College Hospital.

Scope of service

Produce initial outline elemental cost plan for the project works and initial review of cost impact of project risks relative to historical capital costs for new operating theatre projects.

Case study details

Names of parties

- Client: King's College Hospital NHS Foundation Trust
- Contractor: McLaughlin & Harvey
- Project Manager and Lead Consultant: Watts pc.
- Architects: BMJ Architects
- M&E Engineers: WSP
- Engineer: Ross & Partners
- QS: HC QS

Details of project

The client wished to significantly increase the capacity of its existing critical care facilities with the aim of making King's College Hospital a leading trauma and stroke centre in London.

The project involved the creation of additional facilities by adding two new floors containing 24-hour operating theatres and critical care units above the existing building which already comprised two floors with 24 hour operating theatres and a plant room. The works required the removal of the existing roof and provision of a new plant room roof with increased heating, ventilation and air conditioning capacity along with the addition of staff changing and training facilities.

The main challenge of this project was the need to maintain a 24-hour, fully operational critical care unit, whilst adding the two new floors and plant room roofs. In addition, the site was constrained by existing taller buildings around three sides and one of the busiest train commuter routes on the fourth side, which led to limited space for works equipment and materials storage and provided limited options for positioning any cranage. These factors had a significant effect on design risks and buildability issues.

The project was initially split into two phases, with enabling and foundation works being tendered separately from the main superstructure in order for works to commence early. This allowed many technical challenges to be addressed as part of contractor's method statements during the tender period.

The complexity of the project led the team to create a 3D model in order to:

- Bring together the architectural proposals and the structural engineer's frame designs into one 3D model, to assist in providing context to the construction constraints, including working space requirements and the many other factors that impact preliminaries on constrained sites;
- Assist with the visualisation of the engineer's proposals and structural detailing for the complex sections;
- Review the underground arrangements and interactions with existing foundations, building services and working space conditions;
- Assist with identification of cost factors for relocating and protecting existing foundations, services and existing machinery and plant; and
- Enable the high degree of accuracy and detail in the designs for the high-risk enabling works.

Project values

The planned capital expenditure was for £65m over three years, divided in two project phases.

Dates

The HC QS project team provided elemental cost plans and tendering advice between December 2011 and April 2012.

As part of the phase two procurement evaluation process, a benchmark review of costs was carried out in order to compare the hospital project to other projects containing a high proportion of operating theatres. This review highlighted that construction costs for the King's College Hospital project were significantly higher than the upper quartile of all comparable projects used to compile the benchmark data.

While this was expected due to the known constraints of the site and the inherent complexities of building over an existing hospital building, which needed to remain fully operational during construction, the review helped to confirm the key contract cost drivers which were likely to result in significant tender return disparities and the potential for contractual risks during the course of the project. These cost drivers included:

- Greater project risks and extended programme as a result of building over an existing hospital building which had to remain operational during the period of construction;
- Undertaking major building work with significant engineering challenges while surrounded by taller buildings on three sides and a busy rail line on the fourth;
- Building on a site with limited working space, limited storage areas and limited access and turning space for deliveries; and
- Ongoing design development, which could be influenced by the contractor's method statement leading to the need to engage with a contractor prior to the design being completed.

How BIM was used

The structural engineer produced a building frame outline using a BIM design authoring tool. This steelwork model was made available to other members of the design team.

This model, along with the architect's initial concept design drawings (issued in PDF format) at RIBA Plan of Work 2007 Stage C, formed the basis for the creation of a further 3D model by HC QS, which would be used for producing and informing the cost planning process.

During the 3D model creation process by HC QS, particular focus and attention was paid to the external areas, foundations, existing basement structures, groundworks and the structural columns of existing buildings along with the proposed steel members. These sections of the model formed the basis of the quantities data, from which the enabling works tender package was produced.

The model creation process involved adding information labels to model components. These labels were very specific to the taking-off process and contrasted with those created by an architect or engineer who would label the same components with specification details or design-related notes. As a consequence of labelling for taking off, each building component had coding labels that could be linked to a Standard Method of Measurement – in this case SMM7.

As is inevitable in the traditional design process, the information was incomplete. To resolve this, simple block components, or cost assemblies, were added into the model used by the QS so that they could provide quantities for the missing items.

The overall project was still subject to detailed design development and the main mechanical and electrical design consultants had yet to be appointed. This highlights one of the risks with BIM and the model development process, which is that measurement data can be very accurate, but only once the design and details are advanced and modelled to a sufficient level of detail (LOD).

By applying SMM7 labels to the model components they could be grouped into work breakdown structures and work packages for specialist elements. Without this approach to labelling a 3D model for quantity surveying purposes, typical BIM authoring tools would simply provide a schedule of materials. The application of the quantity surveying-specific information labels allowed items to be labelled for construction method items that were required for pricing such as formwork and working space for foundations. By creating a specific 3D model, additional non-design cost components could be added and labelled in line with the required measurement rules.

The process of creating an in-house 3D model for measurement purposes also makes it visually obvious which components have or have not been measured. In addition, the use of 3D model checking tools makes it easier to check the integrity of the components modelled, including any cross-package or component clashes.

The SMM configured quantities were extracted using the standard material scheduling facilities within the BIM authoring software, where the information labelling enabled the resulting quantities, with their relevant coding fields, to be easily transferred to the cost planning tool.

Impact on traditional process

The model creation process in this case took around four weeks to complete, with the final quantity abstraction and structuring process taking a further two to three days.

Once data was extracted from the 3D model, the structure of the cost plan was based on a typical standard reporting format with resource allocations for labour, plant and materials applied to modelled and SMM grouped components.

Tools used

The tools used to support BIM included:

- BIM authoring tool
- Internally created coding libraries with an SMM7 structure
- Spreadsheet, with bespoke macros.

Issues/benefits

This project highlights one of the main issues that affect the production of quantities and cost planning, that of data quality. As with traditional design development, 3D models are created in stages with increasing detail. However, the software tools are only as good as the data. Measuring quantities requires data, and depending on the stage of design development, there may be gaps in the data. Approaches to resolve this missing data requirement can include the creation of cost and/or design assemblies or, as in this case study, simple block components to give approximate measurements.

One of the main outcomes of the modelling process was that junior members of the quantity surveying team became more engaged and productive during the measurement process. Instead of traditional taking off, they were creating an independent and virtual model of the intended structure. This had several benefits:

- While the 3D measurement and taking-off process was supervised, it was quicker and easier to check progress of the 3D model development compared to reviewing and checking a manual working-up process. As a consequence of the process adopted, no manual quantities were produced; quantities were checked using spreadsheets, but there was no manual calculator-based checking of quantities; and the team did not produce a manual report. Overall, the process created more time for the senior members of the team to focus on the cost planning process, including dialogue with specialists and earthworks contractors on options for methods of working under such challenging conditions, and working through buildingup resource allocations for labour, plant and materials.
- The requirement to copy design detailing for the creation of 3D models as part of the taking-off process helped the junior staff to better understand the complex construction detailing and visualise the building and its components in their final context. This was particularly useful to those with limited on-site experience. The nature of the design authoring tools enabled complex arrangements to be built up by relatively junior staff for taking-off purposes.
- The assessment of the site constraints could be better understood with a 3D model when considering storage requirements for off-site materials, waste materials management, limitations for on-site facilities and the logistical complications for removing large volumes of earth. This was an important factor in assessing the likely impact on the preliminaries allocations ahead of the tendering process.
- With 3D modelling being part of the cost planning process, measurements and quantities were always in context, which made checking and reviewing data at any point much easier.

Case Study 2: Network Rail, Waterloo Station, London

Service profile

The main purpose of this commission was to work with the client to identify the potential costs for improving passenger access options at various locations within London's Waterloo Station. Waterloo Station is one of the busiest stations in

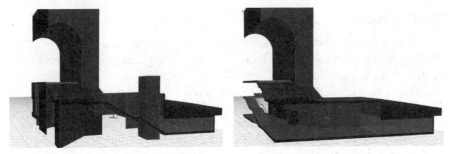

Figure 2.2 Waterloo Station BIM model extract showing existing (left) and proposed (right) layouts.

the world, with usage statistics for April 2011–March 2012 showing 94 million entries and exits made by passengers,[1] with these numbers expected to rise over the coming decades.

The service provided formed part of a wider review of options for the station following the transfer of international services to St Pancras International Station; several outline reviews and feasibility studies were investigated for substantial improvement and remodelling works.

Scope of service

HC QS worked with the Network Rail estimating team to produce 3D models of seven locations where potential improvements had previously been considered. From these 3D models, the main purpose was to:

- Review the relative merits of the options proposed with the emphasis on target cost design;
- Produce quantities for initial cost estimates; and
- Create visualisations to highlight the impact of changes before and after works in order to engage with other Network Rail teams and to provide a basis for presentations to senior personnel to support decision making.

Following client discussions on potential options for station remodelling, seven bills of quantities were produced which formed the basis of outline cost plans. Subsequently the option details and the cost plans were included in wider reviews and feasibility studies for extensive remodelling works at Waterloo Station.

Case study details

Names of parties

- Client: Network Rail-South West
- Project: Waterloo Station, employer lead options development and appraisals
- QS: HC QS

Details of project

In order to explore the range of options being proposed, there was a need to discuss changes that could yield more efficiency, more capacity, improved productivity or generate additional benefits to customers. In order to remove the speculative nature of the options, it was decided to investigate them in some detail utilising 3D Models to significantly reduce the cost risks.

HC QS were commissioned to produce initial budget estimates for pre-feasibility reviews based on significant structural remodelling options, which the client had identified as potential target locations for improving commuter accessibility and safety. The pre-feasibility approach taken was in line with Network Rail's gateway review processes, but with the addition of using 3D models.

At the feasibility stage, it was decided to use the available building plans, fire plans and extensive site survey notes and measurements to develop 3D models of the main features of the building in sufficient detail to enable a broad assessment of the scope of potential redevelopment works.

Project values

Options values: £45m–£75m

Dates

June 2011 to December 2012

How BIM was used

The general approach to producing the 3D models was similar to that used in *Case Study 1*. In order to make budget allocations sufficient for a high-level cost plan and estimate, large blocks of building components were used in order to produce approximate quantities.

This entailed producing detailed component block models, focused around key cost drivers such as staircases, escalators, lifts, floor slabs, walls, arches and passenger foot tunnels in order to provide areas and volumes of key locations. These component blocks were given SMM7 labels and then extracted from the model authoring tools (in an SMM structure); these then formed the basis of the elemental cost plan which was developed using spreadsheets.

Impact on traditional process

A typical approach to early stage options development within a large existing building might require the involvement of building surveyors, engineers and possibly architects in order to put forward sketch plans, which could be used to produce estimates. With the approach adopted, HC QS were able to work directly with the employer to produce sketch plan ideas for options, with development based on relatively straightforward use of 3D modelling techniques. Scale blocks of shapes were produced, which could then be easily reconfigured to produce new layouts. The project quantity surveyors spent time with the employer visiting the existing facilities and reviewing their ideas while collecting survey information and data on key design and structural arrangement issues likely to have a significant cost impact in order to record these as part of the model used for estimating. This had the following advantages:

- The options development process was able to focus on cost-led options to quickly establish the relative value of the different schemes before needing to engage a full design team.
- A traditional process of exploring early stage sketch designs might involve a significant number of stakeholders and, as a consequence, involve the employer in significant cost. By allowing the QS and the employer to review the option costs based on 3D models, the appraisal process was able to engage people using visual context to elemental cost data. By focusing on a cost-led approach to options development, the employer was able to consider the relative merits of taking forward projects based on a better understanding of their financial investment constraints. Employers can, therefore, consider, for example, the relative merits of alteration versus major new works when considering financial viability. While options still require the engagement of engineers and designers, the client is able to establish very early on their needs in the context of their medium- to long-term goals.

The use of 3D models changed the traditional quantity surveying role from one of producing an estimate based on options put forward by others, to the employer and the QS jointly exploring the different options.

Tools used

The tools used to support BIM included:

- BIM authoring tools
- Internally created coding libraries with an SMM7 structure
- Spreadsheet, with bespoke macros.

Issues/benefits

- The use of BIM for estimating and cost planning on this project contributed to a wider review of significant changes and improvements at Waterloo Station. While the project served the purpose of facilitating pre-feasibility cost investigations, the use of 3D models was considered a significant aide to the project team by assisting with the decision process and exposing people to the potential benefits of BIM.

- For this case study, the employer was able to review complex options, seek the views from internal colleagues based on ideas with visual context, and discuss the relative merits and investment costs of the options with senior management, which resulted in time and cost savings on the traditional process.

Case Study 3: Rail Depot Earthworks, Europe

Service profile

The primary role for the HC QS as the project QS was to provide estimating, cost planning and pre-tendering services to the design engineers in order to help identify the optimal design solutions for an inter city rail scheme. The preliminary designs and costs were then used HS2 Ltd as part of the implementation of the wider infrastructure programme compulsory purchase decision making process.

The estimating process included assisting with value for money considerations, identifying cost risks, providing details for the most likely costs of preferred options and assisting with the overall budgets for the section of route being designed by Capita Symonds Ineco for HS2 Limited.

Case study details

Names of parties

- Client: Eurpean Rail Operator
- Engineer: Capita Symonds Ineco (CSI) JV
- Contract Award: Professional Services Contracts, Civil and Structural Design Services
- QS: HC QS (cost consultants to Capita Symonds Ineco JV)

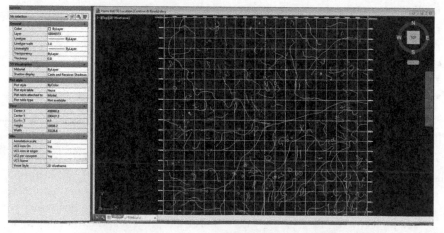

Figure 2.3 Earthworks modelling tool used for HS2 Depot earthworks.

Details of project

As part of an early stage design development, the QS team worked with engineers at Capita to evaluate the most optimal design to meet the infrastructure functional requirements, which were based around train speeds, communities, environmental and value for money considerations. This included such items as locating temporary train sidings, temporary depots, temporary construction compounds and utilities diversions.

The requirement was to assist design teams by providing the costs for the various options being explored.

The HC QS team created 3D models for the temporary depots and sidings for each of the options in order to identify boundaries, scope site areas and overlay levels for various layers of earthwork materials.

These 3D models were then used to produce earthwork quantities, which were transferred to a spreadsheet for cost estimates and cost plans.

Project value

Part of a major rail infrastructure programme, undisclosed budget.

Dates

The case study is based on quantity surveying services provided between October 2012 and December 2013.

How BIM was used

Identifying the volumes of earthworks for depots and sidings was particularly challenging as many of the potential locations for the project impacted to a greater or lesser extent on local communities and local existing infrastructure, such as motorways and trunk roads. As such, the survey work to identify ground levels or ground composition had not been commissioned and would only progress once suitable options were identified. To overcome the lack of local geotechnical information to calculate earthwork quantities and costs, the HC QS project team produced a 3D model from the general design layout data. This model was created by overlaying locations of potential sites and required finished levels onto Ordnance Survey (OS) information obtained in electronic database formats. The 3D model was fixed to the OS data points by reference to road networks and major landmarks sourced from the information available.

From the 3D model, the software used calculated the volume of soil through a series of modelled layers that included the surface levels layer and the lower excavation level layer overlaid onto the existing maps layer to give context to the works by reference to the natural surroundings. Additional information was also introduced into the 3D earthworks models for such items as below ground drainage, petrol interceptors, tunnels and underground structures, as well as the locations of services where they were known.

Using the models was considered the most advantageous and accurate way to calculate the earthworks given the level of detail and information available. The nature of the option evaluation at a variety of sites required an element of flexibility and adaptation to local factors as they become apparent through further investigation. Using the BIM authoring tools enabled changes to boundaries and finished levels to be made and reflected immediately in the overall quantities calculations. This concept, commonly known as parametric design, where changes in one or several areas automatically apply changes to related design components, enables automated remeasurement of work items saving considerable time reworking calculations. This feature of many BIM authoring tools also reduces the potential for errors where project teams are under pressure to regularly update cost estimates and cost plans to meet deadlines.

Tools used

The data and tools used to support BIM included:

- BIM authoring design tool
- Ordnance Survey OS OpenData, mapping data
- Spreadsheet, with bespoke macros.

Issues/benefits

Earthworks can be a major component of cost risks. Many contractors acknowledge that the profitability of a project can sometimes be lost in the ground, primarily because of the inherent risks from working in conditions that are only fully known once excavation commences. In addition to unknown ground conditions, accurately assessing the volume of excavations can also be a difficult task, especially over large areas and on infrastructure projects that are at their early stages of planning and budgeting.

The typical BIM authoring tools used for building or structural design do not always have the required functionality to support earthwork design and extraction of quantities. This may, therefore, require the acquisition and use of additional tools with visualisation capabilities that specialise in assisting with earthworks design and quantification; as well as specialist infrastructure BIM authoring tools beyond the traditional QS toolset.

Undertaking an earthworks exercise starting with 3D data is one of the more beneficial uses of BIM for cost planning. Complex mathematical calculations and dividing up sites using grids can be prone to errors and typically require input from senior members of staff. Utilising 3D models enabled the calculation of the earthwork volumes to a greater degree of accuracy and much faster than would have been possible using traditional measurement methods.

Using BIM for earthworks calculations can be extremely efficient. Complex calculations can be made relatively quickly by software, and sections of the site can be divided up according to design requirements and geotechnical data on ground conditions. The data labelling and coding of areas can then generate quantities that are in SMM form with volumes and areas quickly scheduled. While BIM authoring software used for earthworks calculations has many benefits, it remains of fundamental importance to carry out checks on quantities, boundaries and levels, in order to avoid significant errors, which may result from the use of incorrect levels.

Case Study 4: National Graphene Institute

Service profile

Arcadis undertook pre and post cost management on the highly engineered research building, providing cost modelling and planning services during the design stages.

Case study details

The project was for the provision of a new-build, highly engineered research facility for the development of the University of Manchester's work in the field of graphene.

Figure 2.4 National Graphene Institute.

During the design stages of the project, the team worked under the University's emerging BIM standards and protocols, which cited PAS1192-2:2013 and the AEC (UK) BIM Model File Naming Convention. There was some conflict between these documents as at that time the alignment updates had not occurred. These were subsequently resolved through dialogue.

The design team were not appointed with BIM as a specific deliverable, but as the design stages progressed so did the project's BIM Strategy. It subsequently became a contractual requirement for the contractor to develop a model to comply with the UK Government's directive.

How BIM was used

In this case study, the focus is on the cost planning element of the project. Whilst the design team were not appointed with any BIM deliverables, it was quickly established that given the highly complex nature of the engineering of the project (both structurally and in terms of building services), there could be significant value from using the BIM process.

Arcadis used the outputs from the design teams modelling for a number of purposes including stakeholder visualisations, design efficiency benchmarking, schedule of accommodation validation and cost planning.

In terms of cost planning, the information was used in three ways:

- To verify the quantities already developed within the initial cost plans;
- To offer real-time feedback on design development in terms of key quantity adjustments (and their subsequent commercial impact); and
- To develop detailed procurement packages directly from the model to aid commercial closeout.

At the early stages of the cost planning process, the model was in its infancy. The model was therefore used purely to validate and verify some of the initial allowances for key elements, such as concrete and steel. By engaging directly with the design discipline(s), an understanding was formed of the maturity of the model at the various stages and how it could assist with the cost planning process, as well as understanding what information was missing from the model.

In the light of this knowledge, the model was then used to extract some initial quantity schedules, making allowances for the missing detail to compare against what had been cost planned on a theoretical basis. The importance of this was that it highlighted areas that were potentially deficient, and it also, placed the theoretical cost plan quantities against a more detailed and, potentially, buildable design.

As design development progressed, the various iterations of the model were used to monitor the quantities, such as the volume of concrete. This allowed real time feedback to the design team in terms of design efficiency and affordability.

Once the design had been modelled to a sufficiently detailed stage, the model was used to prepare the pre-tender estimate. This was then issued, without rates, as part of the tender process. The model was also issued at this stage to leverage efficiencies in the tendering process.

Tools used

In order to both view and schedule quantities from the model, a specialist BIM measurement tool was used. Whilst the BIM authoring software would have provided the same outputs, the training required to become competent coupled with the potential to inadvertently change the model meant this option was dismissed.

The software used allowed for simple viewing of the model in a read-only format coupled with the ability to schedule quantities for elements or the whole model.

In terms of the upskilling to use the tools and processes, some external training was undertaken for half a day supported by some internal process workshops to ensure that the entire team working on the project had the skills and understanding to work with the model and produce the required deliverables. These could then be linked into the cost planning platform to give an overarching process from model to cost plan.

Issues/benefits

A number of issues and challenges arising from the process are listed below:

- Understanding the context of the extracted information (i.e. identifying what does not get modelled by the design team);
- Understanding the process of modelling in order to align to the deliverables (i.e. how the levels of detail progress during the design stages);

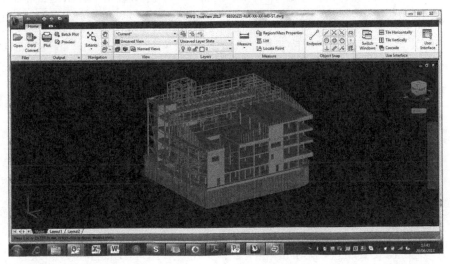

Figure 2.5 3D image of the civil and structural elements.

- Ensuring consistency of modelling to allow elemental alignment (in the absence of a full classification exercise, ensuring that model references remain consistent);
- Aligning the supply chain to model outputs (i.e. whilst the quantity schedules allow pricing, not all supply chain partners are able to view models to understand context); and
- Design liability and intellectual property rights (some members of the design team were unwilling to share their models).

These challenges were overcome in a variety of ways, albeit with the common thread of improved communication. By engaging with the design team and spending time to understand the design process and how the various elements would be developed, the context, process and consistency issues were overcome.

Against these challenges and issues were some key benefits from the process in terms of cost planning. In summary, these were:

- Full understanding of a very complex arrangement by the use of visualisations (these greatly enhanced project team understanding);
- Streamlining the quantity production process by using the model's embedded quantities in lieu of traditional measurement (as noted earlier this occurred in phases, developing from use for verification to quantification);
- Undertaking real-time cost checks against key building quantities to help direct design development to align with the project budget and initial cost estimates;
- Quantity scheduling, allowed reduced periods for procurement package production, offering a more developed design whilst still fitting within the same overall production delivery programme; and
- Reduced risk of quantification errors as model values used rather than double handling.

The real-time feedback to the design team in terms of design efficiency and affordability proved of great benefit. However, whilst this allowed for a better appreciation of the design and the mechanics of quantity scheduling, it highlighted a number of issues when working with early design models. Some of these are:

- Steel member types: until the design is complete, specific members are assumed and so an allowance for additional tonnage is needed.
- Steel member connections: these will not be modelled as part of the process and so will require an allowance.
- Secondary steel allowances: whilst some secondary steel will be modelled for specific areas, such as the roof, a general allowance is still needed.
- Steel member lengths: these do not take account of floor slabs and so in effect give an over measure at the early stages.
- Concrete volumes: until design is complete, depths and thicknesses are illustrative and so an allowance for additional volume is needed.

By utilising the model, the pre-tender estimate work packages were able to be produced in a matter of days as opposed to weeks. The key benefit of this was that it allowed the design to develop as far as possible prior to tendering.

Overall, the use of BIM meant a better design as there was not the need for last-minute savings due to the ongoing real-time monitoring. It also resulted in a more robust commercial position due to the final design being the one that was measured as well as reducing the risk premium for measurement errors.

To Projects

From a project perspective, a number of key benefits were realised from the use of BIM on the scheme. These sat against the common time, cost and quality categories.

Undertaking model linked analysis, the programmed time for construction reduced as the sequencing was optimised. This led to a 10 per cent reduction from the initial planning exercise.

The costs, by a combination of reduced risk and greater design efficiency, also reduced from the initial cost estimate to the final cost position by around 10 per cent. Whilst some of these reductions would have been achieved by conventional methods, the ability to continuously interrogate and intelligently question the model led to continuous efficiencies and a culture of evolution within the design team.

In terms of quality, the project benefited from cost-led design solutions during the design process rather than a cost reduction process at the end of the design phases.

Alongside these key benefits, the following were additional advantages from the process:

- True collaboration by all parties due to the coordinated nature of the required information;
- Open work environment that highlighted exactly where the design sat on a real-time basis;
- Real-time costs achieved by being able to schedule building elements directly from the model and import directly into cost plans, providing immediate feedback on quantum variation;
- Detailed programme analysis for a complex civil engineering package to clearly show missing logic and out of sequence events;
- Reduced contingency allowances due to the robust nature of the information produced and the demonstrated accuracy of the scheduled quantum; and
- Bid deliverables were significantly enhanced by the added value that the model provided, from fly through tours to the inherent coordinated design.

There were some clear benefits achieved by undertaking a model linked programme review and by quantifying key materials directly from the model. The former produced a reduction in the programme of several months. Whilst this

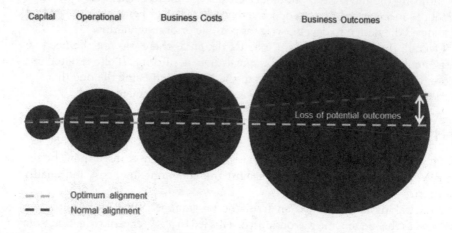

Figure 2.6 Project outcome alignment.

may have been picked up through traditional reviews, the visualisation offered with BIM allowed sequencing to be easily understood and for any missing logic to be quickly captured.

The latter led to a reduction in risk for quantity enhancement/design development. Again, whilst this may have transpired without BIM, its use made this more obvious.

In addition to these two specific items, there is also the inherent benefits that are realised by having a coordinated design. Whilst it is difficult to place a value against this, the principle is that by working in a more collaborative and open environment, projects are better able to achieve their core goals. This then means that, aside from any capital and operational savings, there are wider benefits with regards to the business costs and ultimately the business outcomes.

This can be better explained by reference to the diagram above Figure 2.6, which demonstrates the loss that occurs on most projects in terms of the optimum vs actual alignment of outcomes.

Common issues/benefits

The following outlines a number of generic issues and benefits arising through the use of BIM.

Cost planning using BIM data

At the inception of projects cost parameters and ranges are provided, which are based on particular tolerances of accuracy chosen by experienced clients or the QS. This very early estimation of the most likely project cost can be from +/– 20 per cent rising to +/– 50 per cent for more complex projects.

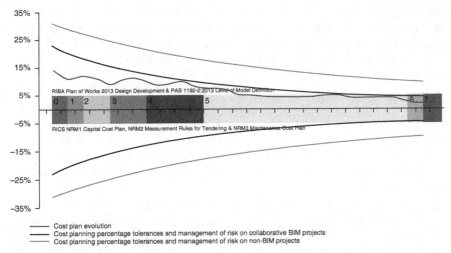

Figure 2.7 Generic issues and benefits arising through the use of BIM.

The underlying benefit of BIM is that it reduces risks and costs by virtual proto-typing of projects before construction works commence, which may also increase the use of out-turn costs as data points for cost benchmarking new projects. This in turn may result in increased accuracy as the cost tolerances at project inception should be much closer, so that, for example, where three-point estimates are produced to give upper, lower and middle cost indicators, the difference between the three points is reduced by comparison to traditional non-BIM-related cost data.

Building up robust BIM cost data will become useful for feasibility studies by those assessing the viability and funding aspects of projects. From contractors seeking loan facilities to banks looking at rating the risk of projects, the consequence of these outline cost ranges is that they establish the budget parameters and expectations, which very often leads to the budgetary constraints within which the entire design delivery team must work.

The services and activities that follow the objective of achieving the budget costs identified during the early cost planning process can very quickly become reliant on the level of project detail available to the QS. The level of available detail is typically based on the status of design development and the level of data accuracy that can be obtained from that information.

BIM *changes to procurement impacting cost planning*

Increasingly, BIM compliant tools are entering the market to assist with some of the budgeting risks by using existing and purchased data to carry out Monte Carlo simulations through to enabling parts of BIM models to be sent to specialist subcontractor for investigative market pricing.

However, some of the biggest changes to cost planning and identifying the most likely project costs are coming from changes being brought forward by the

BIM processes themselves. BIM in context of PAS1192-2:2013 requires a particular method of working to develop virtual prototypes before works begin. This process places contractors at the centre of the design development and coordination process, which will impact the role and function of the QS.

Interactive collaborative cost data

Using technology to deliver greater data accuracy can free up senior staff from data collection and assessment to consider factors that can help to improve cost risk identification and mitigation efforts. This and the increase in productivity that BIM can provide are important value-added benefits of BIM to the quantity surveying process.

Maintaining a cost plan during the early stages of design development through the initial contract mobilisation phase and during the interim payment cycles will require the use of interactive data tools to ensure that cost plans are synchronised with updates to project designs and master programmes. This should mean less time is spent on producing the final output of the absolute detail (i.e. the preparation of detailed measured quantities), something larger firms have been outsourcing, and more time is spent assessing the financial impact of those areas where there is less data or control such as:

- Risk complexities of a project which can define the procurement approach;
- The status of design development and the level of design information available; and
- The many external market factors that can impact on tender prices such as construction inflation.

Reducing costs was one of the defining factors for the introduction of BIM and many of the cost savings objectives are linked to reducing budgetary overruns.

Whether budgets are sufficient or reflective of the most likely outcome can become a material factor for clients and contractors. Many of the legal disputes that can arise on projects have their roots in budgetary constraints that impose a variety of trade-offs and/or incomplete data at the start of projects.

Cost planning around the changes imposed by BIM may become a difficult task during the early transition stages, but an understanding of how BIM impacts risks and, therefore, costs and the degree to which new processes are due to change the timing of cost planning updates will be a key element of using two-way interactive cost data and identifying the level of confidence parameters and weighting to place on evolving cost plans and subsequent recommendations to clients.

Lessons learned

There are a number of lessons to be learnt from these case studies in respect of optimising the cost planning process to obtain the benefit from BIM. There are also wider lessons in respect of project set-up for success from the outset to align BIM deliverables to drive further efficiencies and benefits.

The following provides a brief summary of some of the lessons learned during the design development phase of the case studies featured:

1. While some of the design teams were authoring projects using BIM compliant tools, few made this data available in a reable form and preferred instead to issue paper drawings or PDFs.

    As a result, on some projects it was decided to use the same BIM authoring tools used by the design team to replicate 3D models from the paper and PDF information issued. Whilst the time, effort, technical understanding and creativity to replicate information from scratch can be very resource intensive, replicating the developed designs can be a relatively straightforward 'tracing' process that can be carried out by technician-level staff using the BIM authoring tools.

    Essentially, the process of replicating issued designs using these design tools became the equivalent of taking-off and working-up quantities where design elements and components were labelled within the BIM authoring tools in a manner that would be useful for the various quantity surveying functions and services provided.

2. At its most basic level, BIM produces data and, therefore, early planning into how the data will be generated, to what level of detail and how it will be coded for extraction is essential. Additionally, there is also a requirement to plan how significant volumes of data will be managed and transposed into a useable format.

3. BIM can produce more accurate and versatile data which may be stored in databases and then interrogated, facilitating opportunities to analyse the information in a variety of ways, such as easier benchmarking reporting. However, the source inputs required to obtain the output information can be prone to human error.

4. Junior and technical staff can generate complex 3D models that are used for taking off, but where this process is adopted, senior experienced members of the team are required to oversee the creation process to ensure that what goes in to the modelling process and the allocation of data from models that arrive from third parties are going to eventually be of relevance in the cost plan. Where BIM tools aimed at quantity surveyors are used that can read the data inside models created by others, for example through data standards such as IFC, there is still a requirement to understand how the model has been produced in order to account for components and elements that may be missing, incorrectly labelled or still subject to design development.

5. Typically, BIM authoring tools are structured to suit the particular needs of designers and engineers or for specific elements of design. While typical outputs can allow the generation of schedules of materials, the native input formats for these tools is not usually structured in a way that is immediately helpful for measurement purposes by reference to any standard method of

measurement. There is either a need to work with the design teams to ensure data can be used for cost planning or the QS needs to have a broader understanding of how the models have been put together and to what level of detail, depending on the stage of the project, in order to reclassify the data for specific needs, such as organising information into work breakdown structures for estimating and allocating project on costs such as risk, escalation or preliminaries.

6. While new software tools aimed at cost planning are available, there is still a requirement to understand the limitations of being able to use these tools. For example, 3D BIM models of civil engineering works may provide individual details about the assets to be built or modified, but the design team may not provide information about the context of where those assets are located. In addition, the level of detail within a 3D design may be such that it requires the QS to make extensive use of cost assemblies.

7. The implementation of PAS119-2:2013 requires that the NRM standards are part of the design development and stage sign-off process. There is, therefore, little reason why design data cannot be structured in a way that is informed by measurement rules. This creates technical challenges, initially of where to add measurement codes in models that have competing constraints on the use of data text fields, but it also becomes a cross-discipline coordination challenge where data coding standards are inconsistent.

8. With different BIM authoring tools on the market, each catering for a different need, it has become necessary to make use of simple flat file formats e.g. comma separated files, to enable cross-platform information sharing (COBie). This is of particular importance for data held in legacy cost planning and estimating systems.

This is useful on a number of fronts:

- Data can be extracted from proprietary systems for local use and onward for sharing with clients.
- Reporting to other team members and clients can remain in a readily accessible format.
- The fundamentals of reporting do not need to change in the short to medium term.

9. Quality assurance and compliance with ISO 9001 is an aspect of BIM that can be very challenging when there are different levels of knowledge about the processes that go into the production and generation of information.

Technology doesn't replace expertise and experience; it should enhance it and improve the quality of service delivery.

The outputs of BIM can eventually be formatted into recognisable formats so the information can be benchmarked and checked.

The electronic taking-off process and the generation of data is the main aspect that requires particular attention when using BIM. This is where the quality assurance processes may require updating. One possible solution is to produce detailed process maps of the stages of design development, listing the different tools that were deployed and where additional checking procedures have been introduced in order to signoff the information development.

This type of approach is particularly beneficial when working to standards such as PAS1192-2:2013, which requires the two-way interaction with significant volumes of external data in a number of different formats. Checking that data is complete becomes an onerous task that can only be accomplished electronically. This checking becomes very important, but there are challenges to be overcome such as senior people having the required understanding of the new processes being used and, perhaps more importantly, where they might fail.

There may also be human error issues during the BIM creation process, such as cross-package coordination clashes where BIM models for different disciplines are brought together, or where elements are not typically modelled, resulting in the possibility of items being omitted from pricing.

The need for 'wet signatures' on checked work remains a valuable tool. Whilst these should not be necessary, experience suggests that ISO 9001 procedures will lag BIM adoption. Senior people may also not be directly engaged with the modelling process and still look for information to be supplied in a traditional 'hard copy' format. Some clients also want to see the various senior members physically signoff on deliverables.

It is all too easy to send out electronic data believing that it is correct simply because of the inherent tools within BIM indicating a clash-free design. However, errors can occur within BIM and often they can be hidden in the details and only become apparent later in the process by those that rely on the information at some point during the project lifecycle.

Summary/commentary

Based on practical experience, the overall use of BIM in cost planning has been demonstrated to improve the quality of information being used for cost planning purposes. BIM has also been a way to free up senior quantity surveyors' time away from the traditional and mechanical production of information to enable more time to be devoted to cost analysis, project risk factors and external risk factors all of which have a significant influence on budgets and cost control.

The economic business case for making a change from traditional measurement to direct interrogation of 3D models factored around five early indicators of potential benefits:

1. For most projects, the process of converting PDFs into 3D models, and then extracting the relevant data could be completed in around the same amount of time as a typical traditional take-off. In addition, replicating a 3D model rather than simply extracting measurement information from PDFs, enabled the ability to track changes, effect parametric updates and form the basis of

exploring options with the client. Also, once created, the process of extracting data can be automated and repeated as the revised PDFs are received.

2. The output from the 3D models into a standard method of measurement format to prepare cost plans was easier to check within a 3D model, the data was more flexible, and the output was far more accurate, as it was less prone to human errors inherent in a traditional working-up process.

3. Any changes during the design development period could be processed quickly within the 3D model, making it easier to update quantities, and, therefore, cost plans. Many of the changes to designs would automatically be captured and new dimensions extracted from pre-labelled components, thus avoiding re-measurement. This was found to be one of the primary advantages of working with parametric modelling, where related data is automatically synchronised and updated across all dependent components.

4. Any changes could be visually audited back to any number of previous design iterations, with changes linked to specific times along with the source of the changes.

5. Data created in electronic format could be more easily stored, shared, colour coded and manipulated for any number of uses with an audit trail back to its source. This increased the use of flexible working arrangements, improved retrieval of data and enhanced quality assurance procedures, such as disaster recovery obligations.

While technology is the enabling tool, the premise that technology is *the* driving factor should not be overstated. Since the UK Government launched its Construction Strategy in 2011, the BIM landscape, in terms of technology, has changed rapidly and will inevitably continue to change. The challenges of having to produce 3D models in order to produce quantities is slowly being replaced by cost planning tools that are relatively straightforward and affordable to use, provided 3D models are made available.

The adoption of a defined process using standards such as PAS1192-2:2013 has a greater bearing on the successful outcome of projects and consistency in cost planning than using any specific technology. The defined BIM process dictates *who* does *what* and *when*, and this can impact on procurement choices, risk allocation and level of detail that all need to be factored into cost plans.

The fact that electronic data can be shared between members of the design team and that it is possible to have a two-way automated link to update the data between systems make it even more important to have an experienced QS overseeing the impact on the cost planning and cost advisory contributions.

BIM provides data and assists with producing consistent data records, which can provide confidence to everyone involved, from clients to funders, through to sub-contractors and suppliers. Providing advice and direction on costs is an important aspect of any projects life cycle, and, in particular, in the genesis of a project. It is, therefore, essential, to use any and every tool available, which improves or supports this financial advice.

Had there been the array of BIM documentation that is now seen on a lot of schemes, it is likely that the deliverables and impact of BIM on the case study projects featured would have been much enhanced. Whilst this goes hand in hand with the industry maturity that has since evolved, it is an interesting reflection as the purpose for all of the standards and processes that have been developed is to ensure clarity and consistency. These are absolutes in terms of BIM and its success. As such, by working to the current BIM best practice, it is possible to reduce extended dialogue and bespoke discussions on form and function, and instead to focus on use. This is critical and the only way in which the industry will continue to evolve.

A good example of this is the alignment between the New Rules of Measurement (NRM) and the information model. The BCIS elemental summary was used for the cost plan as this was industry practice at the time. However, now that the NRM has been developed and adopted, it assists in terms of the compatibility, with standard classification forms. Whilst nothing is a perfect fit, there is now an intent for alignment, which helps to reduce the bespoke work elements and further the consistency and clarity. This may be further improved by the adoption of an industry standard classification system (in development at the time of publication).

Note

1   Office of Rail Regulation News, e-mail, 4 April 2013.

References

BSI (2013) *PAS1192-2 Specification for information management for the capital/delivery phase of construction projects using building information modelling*, British Standards Institute, London.
ITGP (2008) *ISO:9001 Quality management systems – requirements*. IT Governance Publishing, Ely, UK.

"YOUR'E GOING TO NEED DEEPER PILES".......

3   Risk and risk management

Steve Pittard and Peter Sell

Introduction

> The biggest risk is not taking any risk. {...} In a world that is changing really quickly, the only strategy that is guaranteed to fail is not taking risks.
>
> Mark Zuckerberg

This chapter considers how and where the use of BIM can impact risk and specifically risk management. All projects contain risk, and their success depends on how (well) we identify, manage and mitigate these risks. Risk can be defined as a combination of the consequences of an event (including changes in circumstances) and the associated likelihood of occurrence (ISO Guide 73, 2013). Cartlidge (2015) further refines this definition as:

> an uncertain event or set of circumstances that should it occur, will have an effect on the achievement of project objectives.

Risk is essentially assessing the probability of something occurring and how much it will cost (if it does) and/or what it would cost to mitigate. This assessment is generally performed at the project level and with the project team assessing the risks and typically capturing them in the form of a risk register, which is structured to identify, assess, prioritise and mitigate the risks. How well (or not) a project performs against its objectives is, therefore, determined by how well the risks, which threaten successful project delivery, are managed.

The Institute of Risk Management (IRM, 2002), defines risk management as:

> a central part of any organisation's strategic management. It is the process whereby organisations methodically address the risks attaching to their activities with the goal of achieving sustained benefit within each activity and across the portfolio of all activities.

The primary aim of risk management is, therefore, to mitigate threats and maximise opportunities. This chapter will consider how BIM could affect the risk profile of a project, together with the range of services which might typically form part of a risk management service and how these services in turn could be influenced by the use of BIM.

Contributors

The editors wish to acknowledge the following for their individual contributions to this chapter:

Julian Downes, Director at J+S Downes (UK) Ltd
Mark Kitching, Partner at EC Harris LLP

Service profile

Risk management may encompass a broad collection of services ranging from strategic (programme/portfolio) to tactical (project). More typically, it is the latter which forms the basis (and consumes the most time and effort) of a risk management service.

Services will generally be tailored to the needs and requirements of the programme or project and may include:

- Project audit
- Project investment analysis and funding support
- Sensitivity analysis (testing sensitivity to changes in input and output variables on the predicted outcome)
- Forensic analysis
- Claims support

However, in construction risk management is essentially about the impact of the risk on cost, time and safety. The core service is likely to be focused around identifying the risks, their potential effect and what allowances or mitigations are required. As with all risk management, the processes applied are based on the following actions: identify, assess, prioritise and mitigate.

Although the scope and nature of services may vary, risk management would typically follow a process similar to that indicated in Figure 3.1 below.

The success of risk management is dependent on the objective implementation of a risk strategy, defined in terms of the agreed response to risk. Typically, these responses would be categorised as follows:

- Risk avoidance (i.e. eliminate, withdraw)
- Risk reduction (i.e. optimise, mitigate)
- Risk transfer (i.e. outsource, insure)
- Risk retention (i.e. accept and budget).

The resulting risk strategy will be influenced by the type of client (i.e. their attitude to risk) and the nature of the project, which in turn will determine the nature of the risk management service required.

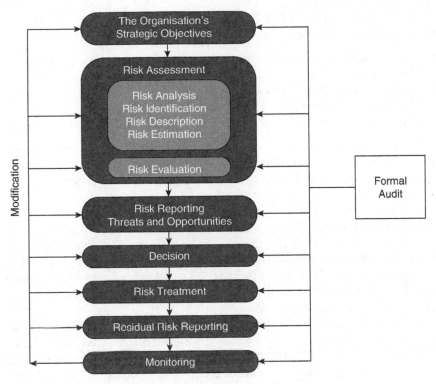

Figure 3.1 The risk management process.
IRM (2002).

Effective risk management requires a systematic approach to the identification, measurement and management of risk. Good risk management procedures ultimately measure the team's confidence level in the project on an ongoing basis (Cullen, 2014).

Scope of service

As indicated above, the range and form of risk management can vary. It may be provided as part of a wider brief forming part of a project controls or project management service. Alternatively, it may be offered as a stand-alone service. In either case, risk management services are often tailored to reflect the needs and requirements of each project or programme and then deployed to provide the necessary strategic and/or tactical support.

At a strategic level, risk management needs to align itself with the employer's (client's) policy, strategy, governance, insurance and controls, so that there is clarity for all parties involved in the project or programme around the risk that the client is prepared to bear, and what mitigating and/or avoidance strategies are acceptable, together with an understanding of the opportunities, of which all parties can take advantage.

At a tactical level, the *RICS New Rules of Measurement, Volume 1* (NRM1; RICS, 2012) now usefully defines risk under the following categories:

- Design development risks (i.e. inadequate project brief, unrealistic programme)
- Construction risks (i.e. access, ground conditions)
- Employer change risks (i.e. changes in scope)
- Employer other risks (i.e. acceleration, postponement)

Whilst quantity surveyors may be keen to promote the strategic aspect of risk, much of their day-to-day service activity is more typically focused on the 'sharp end' of risk management defined here as risk workshops, risk registers, and mitigating actions.

Risk management would typically commence with a risk workshop (RIBA Outline Plan of Work 2007 Stage A/B or digital Plan of Work (dPOW) Stage 0 or 1) from which the required scope of services would be developed and agreed. In a traditional environment, this workshop would identify key project (or programme) guardians for budgets and be based around a review of project information – typically in the form of 2D drawings and related documentation, and often in their hard copy format.

Traditionally, the risk management process is based on a qualitative scoring and assessment methodology, which is, in essence, based on the skill and experience of those involved in the process, where the risks are described, with reference to relevant information about the potential impact of the risks to the project or programme. It is very much a manual process, based on experience, judgement and often time-consuming data collection and analysis. However, this reliance and use of time-consuming manual scoring and assessment is seen as a weakness of the traditional process.

This review and assessment would typically lead to the production and maintenance of a risk register, which is updated and reviewed periodically to reflect changes in project scope, project progress, etc. However, as the frequency of these reviews and reporting processes is often governed by the work involved in collecting the information necessary to provide a basis for assessing the risk profile, this can result in fixed calendar-driven reporting cycles, typically monthly or even quarterly. The risk register is then used to not only identify and assess risks, but also to prioritise and mitigate them.

Quantitative assessment, if it is utilised, would typically come later and is performed at the tactical (project) level, often in conjunction with the project team, who review and assess the risks, with the risk register being subjected to decision tree analysis, cost risk analysis and/or Monte Carlo simulation techniques, which can address both cost and time risk analysis. The outcome of this analysis will be the determination of the correct risk provision that will need to be held, in order to mitigate the cost and schedule risks should they materialise.

Whilst health and safety is categorised by NRM1 as one of a number of construction risks, it is often considered separately, although, in reality, it is just as important when assessing the risks to a successful project outcome.

Historically, health and safety has probably been regarded more typically as a reactive process (e.g. assessing the number of accidents or injuries as a measure of how well the project performed against regional or national statistics). However, as Cook (2015) indicates, clients are becoming increasingly aware of the wider issues which can occur over the life of an asset with many now seeing health and safety as an essential part of the risk management toolset when evaluating investment decisions. This has encouraged some to adopt a more proactive approach to health and safety during construction and throughout the asset life. This proactive approach can help to mitigate a whole host of risks ranging from accidents to the number of lost working days, with the aim of delivering a set of desirable outcomes ranging from improved performance to greater staff retention. Cook (2015) goes on to suggest that health and safety initiatives might, therefore, be implemented as part of a suite of risk management processes, which would include:

- Risk profiling – identify and assess all potential risks, develop mitigation strategies to deal with these.
- Assurance – accurate monitoring, analysis of performance data and control to avoid failings.
- Improvement – identify underlying causes of any failings and instigate corrective measures or new working practices to avoid repeating.

How BIM might impact risk and risk management

Whilst this chapter assesses the impact of BIM on the role and function of a risk management service, it also considers how and where BIM might influence the risk profile of a project or programme, and, therefore, in terms of the wider management of risk, which may, in turn, offer scope to enhance the services being provided.

As risk is largely influenced by both the availability and quality of information, BIM should make the analysis and assessment of risk easier. Risk management can be significantly improved through the availability and use of more structured data. BIM offers a coherent and common source of information promoting the use of common terminology and language reducing or even preventing issues around interpretation where the parties often involved in a traditional environment have their individual and isolated view of the project.

Construction programmes and projects are inherently complex. This complexity carries significant risk that must be carefully managed. The tendency for project teams to work in departmental silos and for there to be a highly separated hierarchy of parties (client, designers, main contractor and sub-contractors) can increase the risk of conflicting information and problems that will have a ripple effect through the project. A BIM environment can mitigate most of these data and information risks, improving effectiveness through better decision making and reducing costs (e.g. elimination of construction waste through collaboration and coordinated design).

In a BIM environment, information should be complete (or at least to the required level of detail or LOD) and be fully coordinated, with the ability to review in 3D and interactively relate with any associated data. There is, therefore, a very real possibility of making use of the vast volume of information to change the processes and practicalities of risk management. With the employment of quantitative analysis and assessment techniques earlier in the life of a project or programme, there will be ready access to significant volumes of structured data very early in the project life all provided through a BIM environment.

At its very simplest, the use of the structured data available in a BIM environment with the 'traditional' (qualitative) risk workshop processes should deliver a better level of analysis because of the completeness of the information available to the participants in the exercise.

However, with the development of BIM, and the structured capture of information which flows from BIM projects, there will be an increased potential for more dynamic assessment of the impacts in terms of time, cost and safety, from various 'risks'. There is also the very real possibility that there could be some auto-analysis, with the potential to visually flag or benchmark key risks for review based on past project performance/experience in a similar way to how costs are benchmarked through industry datasets such as BCIS.

BIM also offers the potential to improve and enhance a post-project review, to assess how well (or not) the project has performed against the expected risk profile.

As noted earlier, traditional risk management is very much a manual process, often involving time-consuming data collection and analysis. With BIM, the data collection could be more dynamic, focusing effort on analysis rather than the collection of data. This may result in a shift in the skill profile, as risk management develops as a specialist, and potentially high-value, service stream.

BIM creates an opportunity to collect performance data to encourage positive behaviours – i.e. to improve health and safety. This also creates an opportunity for the QS to apply their measurement and analysis skills, as measuring the right data not only demonstrates the benefit of a proactive approach to risks, but also provides a catalyst for ongoing performance improvement (see also Chapter 9).

The UK Governments BIM mantra of 'build before you build' clearly offers both the opportunity and potential to reduce and/or eliminate exposure to risk.

How BIM could be used

As already stated, risk management is about the effective identification, assessment, prioritisation and mitigation of risk, which involves the collection and analysis of information. It is the availability and quality of this information that will influence the measure of confidence and certainty. BIM provides the natural environment and ethos to enable this, although this will depend on specifying the data requirements as part of the project information delivery lifecycle.

Table 3.1 is intended to illustrate examples of how and where BIM can improve risk management.

Table 3.1 Examples of how and where BIM can improve risk management.

Potential risk	Potential impact of BIM
Inadequacy or incorrect assessment of the business case	More informed briefing process (ie. PAS 1192–2); Employers Information Requirements (EIR's) & BIM Execution Plan (BEP)
Disputes and claims	Better and more complete information, defined collaborative working framework; reduced/eliminated design development issues
Poor management team	Collaborative working
Poor project information	Improved and more timely information
Design risks	Collaborative and coordinated design; common data environment (CDE)
Construction risks	Integrated programming; virtual construction/ visualisation supporting improved planning, sequencing and buildability
Operational risks	Improved operations and maintenance (O&M) information; Asset Information Model (AIM); Government Soft Landings (GSL)
Re-work	Reduced or eliminated through coordinated design and clash prevention
Health and safety issues	BIM / augmented reality; more informed delivery procedures (see also figures 3.2 and 3.5 below)

Health and safety can be the cause of significant risks in construction which is often accentuated on very complex or technically challenging projects or programmes. The ability to assess, plan and provide a safe working environment is likely to reduce or even eliminate serious risks. Figure 3.2 below illustrates how BIM has been used to both design and inform the construction of some complex temporary supporting works before any work begins on site. The image on the left is that from the digital design, which has been used to communicate the scope, scale and content of the works and compares well with the as-built version shown on the right providing a good example of how BIM can be employed to mitigate risk.

Figure 3.2 Providing an injury-free working environment (Skanska, Paddington Station, Crossrail).

As already indicated, the success of a project or programme is to a great extent dependent on the effective implementation of the risk strategy. The attention given to the risks leads to regular review and assessment of the risks, which in turn leads to the establishment of a risk profile and a risk register. The risk register is used to record, assess, prioritise and mitigate risks. BIM will allow the risk register to be dynamic, as indeed, the management of risks need to be dynamic, which will move the risk management service from a fixed monthly or periodic reporting service to a more interactive and immediate service. BIM offers the potential to make this exercise more dynamic through the collection and analysis of data created or collected as part of the project information delivery process, thereby shifting the emphasis from data/information collection to analysis and assessment, leading to more informed decision making and action. The richer and more dynamic nature of BIM may also reduce the threat of some risks. For example, risks associated with incomplete and/or inconsistent information, which may often result in disputes and/or contract overruns under more traditional project delivery. This is supported by research carried out by Kunz and Gillagan in 2007, where the results of their survey confirmed that the use of BIM lowered the overall risk distribution when compared to projects with a similar contract structure.

The collaborative nature of BIM naturally provides the environment and ethos to support improved risk management, which should involve the entire project team if it is to deliver maximum benefit.

Tools capability

The tools to be utilised within a risk management service may be considered under:

- process and method
- technology.

Risk management is as much about method and process as it is about the tools used to undertake the analysis and assess risk. As noted earlier in this chapter, RICS NRM1 addresses risk as a specific cost centre to be considered when assessing capital cost, encouraging the considered assessment and quantification of risk rather than applying the traditional percentage adjustment for contingencies.

Where risk management is provided as part of a wider project management brief, the IRM reference a number of commonly used standards including ISO 31000:2009 – Risk Management Principles and Guidelines; and ISO 31010:2009 – Risk Management – Risk Assessment Techniques, to guide a process based methodology for assessing and managing risk.

There are a plethora of software tools available to analyse, assess, monitor and report on risk. A number of the most popular tools are add-ons to third party scheduling or cost estimation and management tools with MS Excel often forming the basis for subsequent analysis and reporting i.e. RAG Matrix or Heat Map (see

Figure 3.3 below), which may be further supplemented with the use of dashboards to convey the level of risk exposure (see Figure 3.4).

Clash detection/rendition tools – have the ability to review/eliminate design coordination issues before fabrication and construction, reducing or eliminating impact on the project.

In addition to tools used specifically to support delivery of a risk management service, there are a number of BIM technologies, which can be used to reduce risk on site including the use of augmented reality and Smartboards. Figure 3.5 shows how Smartboard technology can be used to review and manage potential risks with members of the project team. Delivering access to the project model as part of the CDE offers obvious advantages over the more traditional static information in the form of 2D drawings and paper documents.

Whilst there are some examples of these tools being used as part of a fully integrated BIM solution (as seen in Figure 3.5), this is far from being typical, and even where they are used, they are often employed in isolation. It is, therefore, likely that there will be convergence around some of the common tools as the industry becomes more familiar with the availability and use of a richer data set, creating the potential for wider integration and analysis to improve and enhance risk management services.

Issues/benefits

BIM will arguably create more pull for risk management and particularly early strategic engagement as clients require this input and expertise to develop their EIRs.

As with many aspects of BIM information, it is all largely about how well the data is structured in relation to the tasks at hand, and effective risk management will depend very much on the classification systems employed.

Impact / Probability	Negligible	Minor	Moderate	Significant	Severe
81-100%	Low Risk	Moderate Risk	High Risk	Extreme Risk	Extreme Risk
61-80%	Minimum Risk	Low Risk	Moderate Risk	High Risk	Extreme Risk
41-60%	Minimum Risk	Low Risk	Moderate Risk	High Risk	High Risk
21-40%	Minimum Risk	Low Risk	Low Risk	Moderate Risk	High Risk
0-20%	Minimum Risk	Minimum Risk	Low Risk	Moderate Risk	High Risk

Figure 3.3 Example of a RAG Matrix or Heat Map.

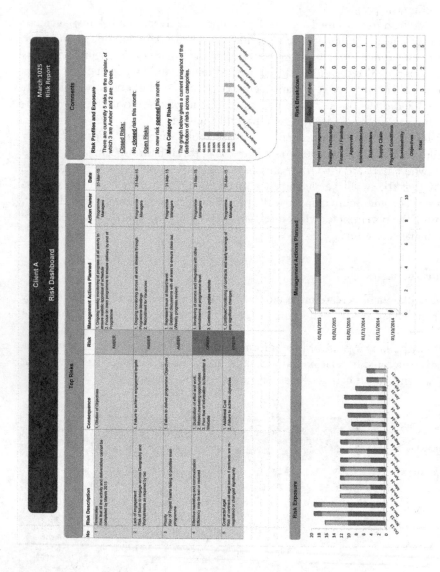

Figure 3.4 Example of a risk exposure dashboard.

Figure 3.5 Use of Smartboards (Skanska, Paddington Station, Crossrail).

Traditional risk analysis is typically a bottom-up process (i.e. analysing and aggregating the detail). BIM should facilitate the ability to model from historic data; albeit, this will depend on the quality and consistency in classification. This may offer opportunities to develop risk databases. However, there will be issues around model and data ownership, liability and security, with the need to restrict availability for general/public analysis (i.e. information from past projects will be time sensitive and may form the basis of disputes, or be embargoed for other reasons).

Collecting the right data to create the necessary feedback loop for communication with others is key. This will require a more open approach to the exchange and sharing of information, which may meet with some resistance until the industry becomes more familiar with sharing information for mutual benefit.

The inherent use of diverse breakdown structures to classify different aspects of the same project can be an issue for risk management i.e. Cost Breakdown Structures; Work Breakdown Structures and Risk Breakdown Structures/ Mitigation Breakdown Structures do not easily align. To obtain the benefits that BIM promises in this area it is clear that time needs to be devoted in the very early stages of a project to developing the breakdown structures to enable the analysis of the data in a hierarchical risk breakdown structure.

Despite the obvious motivation for aligning, this is likely to take more time as the industry begins to address risk as a standard part of the cost and time assessment.

It is perhaps also worth noting that BIM introduces a few new risk areas to be managed:

• Legal liability and responsibility along with the introduction of new and unfamiliar forms of contract and procurement.

- Technology both in terms of the issues relating to information exchange and interoperability as well as the complexities of using large integrated systems.
- Increased risks associated with faster response times (less thinking time!).
- The inherent complexities combined with the (yet to be) universally understood nature of BIM may lead to incorrect and/or inconsistent assumptions in terms of the nature and quality of information being provided i.e. the failure to challenge or validate the data as the basis for decision making.
- Overcoming the legacy silo mindset!

In addition, the use of BIM-enabled mobile devices in the workplace may also introduce some new health and safety risks (i.e. using near open excavations, machinery, etc), which will need to be managed as part of the health and safety regime.

Whilst some of these areas are dealt with in more detail in other chapters, the integrated concept of BIM can blur the level of responsibility so much that risk and liability might actually be increased. One of the most effective ways of dealing with (mitigating) these risks is through the use of collaborative, integrated project delivery contracts where the risks of using BIM are shared.

Key benefits

To the project/programme

By its very nature the implementation of BIM will have a positive effect on the risk profile of those projects and programmes where it is used. Some of the key areas where BIM will have an effect are:

- Effective BIM implementation helps avoid design alterations, reducing risk and time in reworking the detailed solutions.
- BIM allows the project team to effectively look into the future and experience how a facility will work before completion (i.e. 'build before you build'). This helps to reduce uncertainty, increase confidence and thereby reduce risk.
- BIM should reduce errors in the design process, with the standardisation of information structure and the collaborative processes that BIM engenders and hence a reduction in risk; for example, through the use of standards and protocols such as PAS1192-2:2013 and the CIC BIM Protocol (2013). The Ministry of Justice (MoJ) have indicated that the early identification and mitigation of a significant error on just one of their projects delivered in compliance with BIM Level 2 equated to a saving equal to the departmental cost of implementing BIM.
- BIM improves the project team's ability to understand and resolve the detail earlier in the design process and thereby reduce the construction risk.

As already outlined in this chapter, BIM also enables construction teams to plan work more safely (i.e. excavations, proximity of services, construction traffic flow in close proximity to open earthworks, etc.) as well as highlighting potential areas of safety risks and providing assurance to the construction team that the planned working sequence and methodology provide a safe method of working, thereby mitigating the risks associated with complex construction sequencing or interfaces.

To the businesses

The collaborative nature of BIM processes and a focus on the dynamic integration of scheduling offers the opportunity to reduce risks inherent in highly complex projects. Risk reduction is an important factor on any project as it enables the client to have assurance or improved confidence that the project can be built on time and within budget.

Figure 3.6 below shows the risk reduction projections calculated for modelling deployed across a major infrastructure project. A detailed analysis of the project indicated a risk reduction of at least 5 per cent, which equated to reducing the risk profile by over £8m.

The benefits of schedule-integrated models are reinforced by a comment from a Project Manager on a major infrastructure programme:

> This exercise helped identify and resolve clashes between works. The tool is a highly visual and collaborative way to ensure that we have confidence

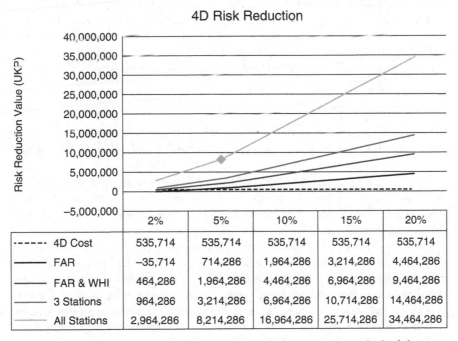

	2%	5%	10%	15%	20%
------ 4D Cost	535,714	535,714	535,714	535,714	535,714
——— FAR	-35,714	714,286	1,964,286	3,214,286	4,464,286
——— FAR & WHI	464,286	1,964,286	4,464,286	6,964,286	9,464,286
——— 3 Stations	964,286	3,214,286	6,964,286	10,714,286	14,464,286
——— All Stations	2,964,286	8,214,286	16,964,286	25,714,286	34,464,286

Figure 3.6 Chart indicating risk reduction through the integration of scheduling tools with BIM.

in the 'buildability' of the works in a tight schedule, and considering the numerous scope packages, physical interfaces, overlapping works, and stake-holders that are typical to a project.

To others

In addition to the above and looking to the future, the on-going development and utilisation of BIM will allow risk management and risk identification pro-cesses to develop in the future on the basis of the enhanced information capture from BIM. For example, and as indicated earlier in this chapter, BIM may offer the opportunity to benchmark risks in a similar way to that currently used to benchmark capital costs. This could comprise an array of metrics upon which to assess performance across a range of selected risk categories including health and safety.

The adoption of a more unified Risk/Mitigation Breakdown Structure (mapped to the Cost Breakdown Structure and/or Work Breakdown Structure), would enable a greater degree of certainty and accuracy in the analysis and assessment of risks.

The lack of available reliable quality structured data may severely hamper the ability to overlay data on other data, based on common referencing and data hierarchies, to provide an informed view of the potential risk(s).

The improved information base created with BIM will also provide the oppor-tunity to build historical risk profiles based on experience and lessons learned, which should allow for improved and informed management of future projects.

It is likely that the use of BIM could lead to a reduction in the risk allowance held to make provision for the unknowns, as there should be increased certainty from the use of the BIM methodology. This in turn may also lead to reduced finance charges, reflecting the reduced risk profile of BIM projects.

Summary/commentary

BIM offers the potential to reduce risk beyond project delivery, where the predict-ability of building performance and operation can be greatly improved (Azhar, 2011).

BIM has the potential to unlock a range of opportunities services to support improved risk management. Construction projects, large or small, are all about managing risk and this is a key driver in any investment decision. The lower the risk the higher the reward!

Clients with large programmes will start to build data sets to measure perfor-mance over time (i.e. scheduled versus actual progress linked to the model) to track the forecasting accuracy of their contractors and related supply chain. This will then enable the ability to benchmark performance in order to assess the risk of using particular organisations, which will in turn impact capital project finance underwriting along with the pre-selection of contractors for major programmes.

It is clear that BIM increases the value of risk management and the scope to address risk. However, will this necessarily enhance the value of the QS? To analyse this, there needs to be consideration of how much of risk management is about data/information collection as opposed to analysis and the skill of mitigation. BIM certainly offers the capacity to significantly reduce – and maybe even eliminate – the former, as well as support the latter, so this will logically lead to a shift in emphasis and service focus if the QS is to realise the opportunities offered by BIM.

References

Azhar, S. (2011) 'Building Information Modeling (BIM): trends, benefits, risks, and challenges for the AEC industry', *Leadership and Management in Engineering*, 241–252, July.

Cartlidge, D. (2015) *Construction Project Manager's Pocket Book*, Routledge, Abingdon.

CIC (2013) *BIM Protocol*, Construction Industry Council, London, UK.

Cook, M. (2015) 'From red tape to risk management', *RICS Construction Journal*, 11–12, April/May.

Cullen, S. (2014) 'Risk management'. The Whole Building Design Guide: WBDG. Online. Available HTTP: <www.wbdg.org/project/riskmanage.php>, accessed May 2015.

Infrastructure Risk Group (2013) *Leading Practice and Improvement*: Report from the Infrastructure Risk Group, London.

IRC and IRM (2013) 'Managing cost risk & uncertainty in infrastructure projects'. IRM. Online. Available at <https://www.theirm.org/media/654694/IRM-REPORTLRV2. pdf>. Accessed 21 July, 2015.

IRM (2002) 'A risk management standard'. The Institute of Risk Management, London. Online. Available HTTP: <www.theirm.org>.

ISO Guide 73 (2013) *Risk Management – Vocabulary*, British Standards Institute, London.

Kunz, J. and Gillagan, B. (2007) 'VDC use in 2007: significant value, dramatic growth and apparent business opportunity', Online. Available at <cife.stanford.edu/sites/default/files/TR171.pdf>, accessed July 2015.

RIBA (2007) *Outline Plan of Work*, RIBA, London.

RICS (2012) *New Rules of Measurement 1: Order of cost estimating and cost planning for capital building works*, Second Edition, RICS, London

Succar, B. (2009) 'Building Information Modelling Framework: a research and delivery foundation for industry stakeholders', *Automation in Construction*, 18, 357–375.

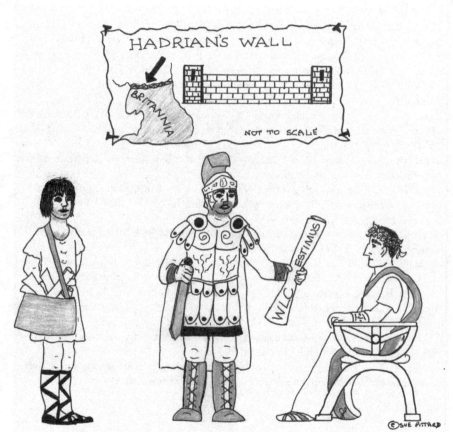

'HE SAYS IT SHOULD BE GOOD FOR AT LEAST FIFTY YEARS'

4 Whole Life Costing

Yemi Akinwonmi

Introduction

Building Information Modelling (BIM) is beginning to change the way we build, the way buildings look, how they function and the way buildings are maintained and managed (Godager, 2011). The Institute of Civil Engineering, ICE (2012) describes it as the management of information through the whole life cycle of a built asset and in the context of whole life costing. BIM can ensure a thorough life cycle analysis, service life planning and more solid life cycle optimisations of the design and use of the buildings (Hallberg and Tarandi, 2011).

Brief outline of content

This chapter seeks to consider a case study demonstrating how BIM can be applied in the area of whole life costing. It appraises the approach taken in the application of BIM, the processes and tools utilised, the challenges faced and opportunities for further development. An example schools project has been used to demonstrate the capability and value of the BIM offering to clients and projects in the area of whole life costing.

Author biography

Yemi Akinwonmi is a built asset consultant with a strong quantity surveying background having experience across the life cycle spectrum of construction and engineering projects. He has a bachelor's degree in quantity surveying, a master's degree in facilities management and asset performance, and several professional affiliations including the Royal Institution of Chartered Surveyors, RICS and the British Institute of Facilities Management, BIFM.

Service profile

Whole life costing is an economic method of project evaluation in which all costs arising from owning, operating, maintaining and ultimately disposing of an asset are considered in the evaluation of design decisions to achieve long-term value for money, defined here as the optimum combination of whole life cost and quality to meet the user's requirement. 'Awarding contracts on the basis of lowest price tendered for construction works does not necessarily provide value for money, it is the long-term value over the life of the asset that provides a much more reliable indicator. It is the relationship between long-term costs and the benefit achieved by clients that represents value for money' (OGC, 2003).

'Whole Life Costing (WLC)' is defined in BSI ISO 15686-5:2008 as a

> methodology for systematic economic consideration of all significant and relevant initial and future costs and benefits of an asset, throughout its life cycle, while fulfilling the performance requirements over a period of analysis, as defined in the agreed scope.

Kelly and Hunter (2009) describe it as:

> an exercise in which the capital cost of the project and all relevant future costs are made explicit and used either as a basis for a cash flow projection over a given period of time, or used in an option appraisal exercise to evaluate various solutions to a given design problem.

A practical outworking of the WLC approach is to consider 'cost-in-use' during the design process by undertaking life cycle appraisals of the design options to aid the design process by ascertaining the life cycle costs. These studies inform material/component specification, space layout, location of services and key facilities, and which are best facilitated via close working relationships with design teams engaged with facilities management (FM) practitioners in order to achieve the necessary balance between capital costs and longevity over the life cycle of the assets.

Generally, in a traditional environment, life cycle costing and analysis are carried out using spreadsheet-based models which provide outputs to give an indication of the scope of the life cycle maintenance/replacement frequencies on building or facility elements that would be required over a given and specified period of time. It is important to note that there are numerous factors which interact to determine the life of building components; and life cycle models are only intended to identify a suitable overall fund, along with an indication of when individual items are likely to need to be replaced. The BS ISO 15686 provides a standardised approach to life cycle costing that covers the inclusion of maintenance costs in a life cycle costing exercise which in turn refers to any planned and reactive maintenance activities such as inspections and servicing the whole or part of the building or facility assets.

Life cycle replacement cost estimates and profiles are based upon an assessment of the expected service life of each asset/component and the likely replacement cost at the end of that service life. Service life expectancies of building or facility assets can be obtained by published data from industry sources such as CIBSE, BMI and BCIS along with manufacturer's product and warranty data. However, it is important that practical operational considerations such as location, nature of use, occupancy patterns etc. are also taken into account. The industry standard approach involves building the models using construction cost/design information and historical data, whilst considering specific site exposure and operational conditions. Hence, effective collaboration is required between relevant members of the design team and other professional advisors, including a high degree of interface between Architects, Quantity Surveyors and M&E consultants at different stages of design development.

The following provides a summary analysis of a typical WLC service stream mapped to the Digital Plan of Work (dPOW):

- Appraisal and Design Brief Stage (dPOW stage 1) – budget/benchmark life cycle plans are produced to give an indication of potential life cycle cost position.
- Concept Design, Design Development and Technical Design (dPOW stages 2 and 3) – elemental, sub-elemental and component level life cycle costings are produced. In addition, during these design phases Option Appraisal exercises are carried out to explore which design solutions deliver the potential of best value for money, along with informing BREEAM life cycle assessments where appropriate.
- Production Information and Tender Documentation (dPOW stage 4) – final life cycle cost checks and analyses are carried out.
- Construction (dPOW stage 5) – on-going life cycle cost checks can continue during the construction phases to identify the impact of variations/changes to design and specification. At this point, BREEAM assessments and reports are also reviewed.

Life cycle models are also being increasingly used as a cash flow monitoring and management tool during the operational phases of projects especially under complex contractual arrangements, such us Private Finance Initiative (PFI) as it is commonly called in the UK.

Case study details

The case study is based on a data-rich 3D model with a focus on a particular work stream that relates to whole life costing, which the author was responsible for leading.

The content of the case study was based on a 'sample school project' commissioned by an in-house multi-disciplinary team accountable for the development of the BIM capability and its implementation across a global engineering enterprise.

The sample school project was investigated to explore and demonstrate the capability and value of BIM to clients and projects in the area of life cycle costing. There was a key focus on achieving increased clarity on long-term operational performance expectations and the consequences of design-stage decisions

Figure 4.1 Featured case study image.

over life cycle performance and cost. In addition, the sample project attempted to demonstrate BIM as a driver for reducing project start-up costs, through the availability of better information at the beginning of a project and better life cycle management resulting from the readily available consolidated design and construction information/data sets. The research also acted as a potential test of the interoperability between various tools and software applications chosen to support the BIM common data environment (CDE). The project also explored how to maximise the opportunities for the collaboration of different roles and functions within the design and delivery teams.

The following provides a summary of how the WLC service was delivered around BIM for the sample school project.

How BIM could be used

There are significant limitations with 2D drawings as they lack the rich three-dimensional (3D) context, that the quantity surveyor (QS) needs in order to identify important cost-sensitive design features (Shen and Issa, 2010). BIM tools have helped the QS visualise real-world conditions through a virtual three-dimensional construction of the building (Sylvester and Dietrich, 2010). Whilst there is no shortage of software applications offering the measurement, pricing and programming functionality required to estimate capital construction costs, limitations within the WLC equivalent toolset (at the time of the project), restricted the QS gaining a similar life cycle cost perspective.

Within the context of life cycle costing (LCC), the sample school project set out to achieve the creation of a 'data rich' 3D model embedded with building component life expectancy data, via a 3D object library containing multiple data/attributes, to facilitate life cycle options appraisals.

The 3D model was linked with other relational cost and specification data-bases, along with a scheduling programme. This linkage enabled the calculation of the life cycle costs, and, via the 3D model, gave a visualisation of the life cycle replacement works. The resulting data rich model then had the ability to show the impact on life cycle cost following changes in specification or different objects in the 3D model. Furthermore, this then allowed a visual demonstration of the benefits of life cycle options appraisals early in the design process. This visualisation helped designers and informed design decisions, specifications, and, potential solutions, through the developmental stages. The use of visualisations also helped the project and wider team understand the methodology and value of life cycle costing through its application.

Figures 4.2 and 4.3 provide examples of how WLC can be integrated with BIM.

Figure 4.2 illustrates the how specifications can be changed for a BIM-enabled life cycle costing exercise using a section of a building floor plate.

Figure 4.3a and 4.3b indicate how WLC may be incorporated as part of an integrated BIM solution.

Impact on traditional processes

Traditionally, a significant part of the life cycle cost analysis is based around individual components and has typically been delivered by the QS in isolation, which in essence, manifests as an exercise at the completion of the design stage,

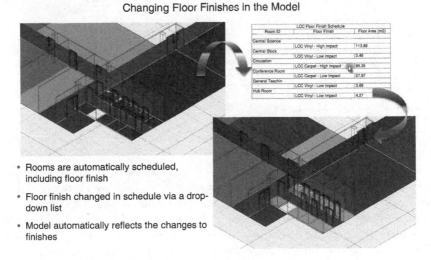

Figure 4.2 BIM-enabled life cycle costing exercise.

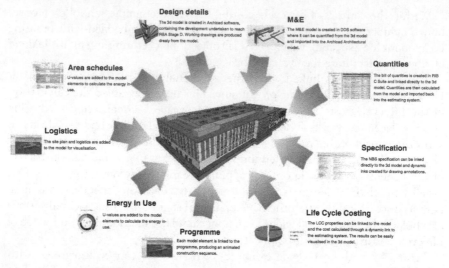

Figure 4.3a WLC BIM inputs.

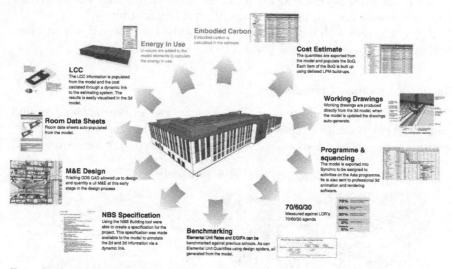

Figure 4.3b WLC BIM outputs.

with the results used to assist the client to understand the on-going cost of maintaining the building or facility.

In a BIM environment, with the life cycle data stored in the working 3D model as object attributes, the data is immediately available to all relevant parties involved in the project. This immediacy fosters increased engagement and interest from all other parties engaged in the project, and enables the life cycle assessment process to be carried out collaboratively and dynamically throughout design development. The life cycle costing outputs are driven via data in a central database specific to the project, which helps to achieve better consistency, eliminating ambiguity and providing the opportunity for a quick review of the scheme, hence, maximising value through manipulation of the model.

As the data can be accessed directly from the model, there is less need to replicate it for use with spreadsheets, which in itself should result in a better appreciation of the life cycle costing process by the project team. The centralisation of the data is a huge step towards a holistic approach for project data management as rolled out by BSRIA, and this is supported by the Government Soft Landings (GSL) frameworks. In addition, this ensures that a CDE is established and maintained so that the data handed over is in the correct format and contain the right data fields.

This enriched data pool will be further enhanced by the adoption and implementation of a unified classification system, which forms part of the UK Government's BIM strategy and roadmap required to meet its 2016 target. Historically, WLC has lacked the universally accepted classification and approach to analysing asset life cycle costs, and it is hoped that WLC will benefit in the same way that cost planning benefited from the introduction of the widely adopted BCIS elemental cost breakdown now embodied in *RICS New Rules of Measurement, Volume 1* (NRM1). The lack of any real standardised approach to WLC has also inhibited the availability of any industry standard benchmarking data, as the format and structure has not lent itself easily to universal comparison and analysis. The introduction of the *RICS New Rules of Measurement for Building Maintenance, Volume 3* (NRM3), coupled with industry-wide adoption of a unified classification system, is likely to lead to a transformation in terms of both the demand and delivery of services around WLC. The creation and availability of consistent historical life cycle cost data for analysis and benchmarking is also likely to see these services increase in value as the industry seeks to respond to the UK government's Construction 2025 strategy calling for a 33 per cent reduction in whole life costs.

Tools used

Fundamental to the achievement of effective BIM implementation in the area of life cycle costing is collaboration between all relevant project stakeholders to ensure that the necessary data exchange protocols are established and processes are in place to enable project integration.

Various tools were used in the deployment of BIM on the sample school project, including CAD-based applications which were used to create and manage the data rich 3D model linked to other relational database-driven software, which brought the intelligence of cost estimation and specification generation. Other proprietary interfacing applications were introduced to ensure that multiple measurement attributes were attached to the relevant objects in the 3D model. The model was also linked to scheduling software to facilitate a visualisation of the sequencing of life cycle works, which enabled the production of an animated sequence of life cycle works to be undertaken at defined time intervals. The use of a high-resolution graphics software application enabled the visualisation of the life cycle works sequencing to be easily and readily shared with the client and the project team. This, visualisation whilst being the output of detailed life cycle analysis in itself, brought greater clarity and understanding to the team of the scale, scope and frequency of the life cycle works.

The sample school project found that undertaking the life cycle cost calculations in the 3D model was not practical, or indeed possible. As such, in order to overcome the limitation of the 3D model, objects/elements in the 3D model were linked to a cost estimation database and life cycle model, where all the complex calculations were undertaken.

In the first instance, a full bill of quantities (BQ) was created in a database driven estimating system, which was linked directly to the 3D model. This allowed the required quantities to be derived from the model for use with the estimating software. Each item of the BQ was built up using detailed labour, plant and material data. By using an associated specification tool, it was possible to determine the specification of elements/key components of the project. This information was in turn made available via a dynamic link, to the models.

To facilitate options appraisals, one of the key benefits of life cycle cost analysis, the element specification was changed directly in the 3D object library, which was embedded in the data rich CAD model.

In practical terms, this involved the use of a drop down menu in an activity schedule derived by a tool, which was used to associate data and properties with the appropriate 3D model objects. Changes in specification and quantities within the 3D Model were then output into the cost database via an automated link. The relevant sections of the BQ were populated and updated to produce cost breakdowns and annual cost allocations. The resulting changes in cost calculations were then automated and selected outputs exported back into the 3D model. The final step in the process was to adjust the life cycle works programme. This task had to be undertaken manually using the scheduling software.

Utilising a full BQ proved to be quite cumbersome and resulted in too much information being visualised when there were significant life cycle replacements and other maintenance activities to be carried out. It was subsequently decided to work through the models in sub-elemental sections, or, in the case of finished elements, to compartmentalise the 3D model using floor plates or zones, to overcome the data volume issue.

There were a number of additional outputs that were achieved, from the use of the processes and procedures outlined above, namely:

- Area schedules created from the 3D model – these were used to inform the potential maintenance requirements as the functional spaces were better appreciated;
- U-values were added to the model for high-level energy-in-use calculations using a set of assumed parameters; and,
- Outputs were benchmarked in cost per square metre.

All of the above were generated directly from the 3D model.

Issues/benefits

Summary of issues and problems to be overcome

One of the major limitations, as explained by Shen et al. (2007), is that current IFC models are not semantically rich enough to cover the entire construction process data and job conditions. This fact, coupled with the realisation, as detailed by Sylvester and Dietrich (2010), that it may require several software packages to create viable and usable results, show that there are restrictions in interacting within the BIM interface to produce a life cycle costs assessment within the 3D model.

An alternative to the predefined-data-model approach is to allow the QS/cost consultant to apply their own domain-specific judgments to the design features with the assistance of 3D visualisation and quantity data from BIM models (Shen and Issa, 2010). This was the solution adopted for the sample school project. However, it is clear that to successfully deliver the required integration and automation requires that the QS determines for each given situation, which software applications and data exchange tools are required for the specific CDE employed.

The technology challenge and limitations of current software should not be underestimated. There are also significant issues with software interoperability, although this is now being addressed through engagement by the industry with software developers, who are, where possible, working together to develop solutions to help achieve seamless data transfer between domains and disciplines. Whilst the application of BIM in this manner increased the awareness and interests from other members of the design team, there still exists a major skills gap in the industry of those who are able to work and develop BIM data exchange protocols.

There is a predominant negative attitude towards the widespread adoption of working with BIM in the life cycle cost arena, which will only be overcome by training and the enforcement of BIM Execution Plans.

A summary of the key benefits that could result from the use of BIM

To projects

BIM enables 3D visualisation of the methodology and impact of life cycle assessment to provoke discussion and optioneering with the design team. This visualisation and data hosting helps to create an environment, which should bring better understanding of how the building or facility will potentially operate post completion, creating an enabling platform for facilities and asset management which is impossible to achieve in a spreadsheet-based life cycle model (see also Chapter 11).

One of the key benefits of BIM to a project is effective data management and increased engagement with project stakeholders. On the case study project, whilst it was the first time life cycle costs were linked directly to the 3D model,

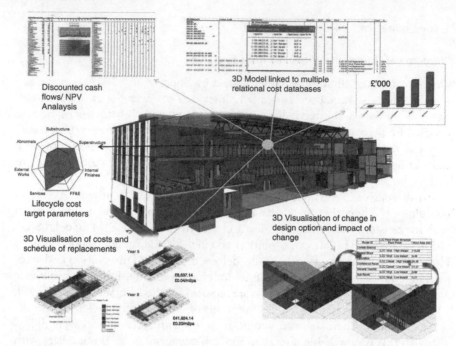

Figure 4.4 Various LCC outputs that can be drawn from the 3D model.

cost database and scheduling application, the major benefit was in faster option-eering and a better understanding of life cycle management, which resulted from the consolidated design and construction information being made available through a single source (CDE).

The use of the CDE increased the confidence levels in the prediction of the long-term asset replacement and cash flow forecasting of the project, as well as creating an opportunity to drive efficiencies in the replacement programming and procurement of the asset.

To businesses

The adoption of BIM provides a wide range of broader benefits to businesses as information can be value added, shared and reused for different purposes. With the push for the use of GSLs in the UK, businesses will be encouraged to ensure better briefing, design, handover and asset performance in-use data is maintained, which should enable the transition of information from inception through design and construction into occupation. This will, in turn, allow the optimisation of operational costs, energy consumption and longevity of built environment assets.

In the case study, the use of BIM meant that building proposals could be rigorously analysed at a strategic level, simulations could be performed quickly and performance benchmarked, enabling improved and innovative solutions.

At an early stage of the project, when price, but perhaps not detailed design, was being considered, BIM provided the ability to give a comprehensive schedule of assumptions, upon which the project was based and performance function requirements defined within the specification. This reduced the risk in terms of future escalation of costs due to changing quality requirements.

The use of BIM also enabled the automated assembly of efficient building components, which meant that the digital product data could be exploited in downstream processes and used for manufacturing/assembling of the building systems/components. This means of working also drives standardisation of products used, which in turn allows for better buying power and quicker and easier evaluation of competing options in procurement. In the sample school project, the long-term costs and value was taken more seriously and there was an improved awareness of life cycle costs on a much broader scale.

The value of operational cost input into the design process is much more feasible, and can be utilised across a portfolio. The availability of the data in useable formats means that benchmarking and continuous improvement can be facilitated, enabling design lessons learned to be passed from project to project. All of this should result in less waste, lower emissions and better value, not only in the operational phase, but also in the construction phase, which will enable businesses to be more sustainable in their operation, and facilitate the creation of a highly valuable knowledge base for future schemes.

To others

In addition to all the inherent benefits that BIM can bring to projects and businesses, there is also the value it can add to end-users via effective integration with designers, principal contractors and Mechanical Electrical and Plumbing (MEP) commissioning managers. Digital documentation solutions can be used to generate operations and maintenance (O&M) information, building (or asset) user guides, tenant guides, health and safety files, project completion files, BREEAM user guides and Energy logbooks. Asset registers and planned preventative maintenance (PPM) schedules can also be produced as an output for the FM helpdesk service and its Computer-Aided FM (CAFM) system to enable long-term asset management (see also Chapter 11, which deals with FM in more detail).

The sample school project enabled others to understand, through visualization, the outputs and options available. This in turn brought about better clarity and an understanding of the areas that required improvement, especially in long-term in-service performance expectations and the consequences of design-stage decisions over life cycle performance and cost.

As a result of the use of BIM, a data rich 3D model was produced, which could be used to engage with multiple stakeholders, in order to help them to understand the space, (i.e. in terms of how to use, service and maintain it). The model also provided a visual interface and foundation for other systems such as CAFM, Building Management Systems (BMS) and Energy Monitoring Systems (EMS) (see Chapter 11 for more detail on these systems).

The BIM process also encouraged greater collaboration, as all parties were able to engage in collaborative 3D viewing sessions, which improved communications and trust between stakeholders. This also enabled a broader level of knowledge sharing and rapid decision making early in the process.

It is salient to note that whilst BIM has been known to deliver a number of tangible and measurable benefits through the design and construction phases of projects, its value in the post-construction/operational phases of projects is only just being explored. From an asset life cycle perspective, quicker and more effective optioneering with improved engagement within the design team is an easy and obvious win.

For example, structured data (i.e. COBie) can be set up to help with appraisals such as those required for BREEAM, and in a perfect world, the information from manufacturers could be collected and then used to inform the design selection.

There is no doubt that in the operational phase, having a complete, accurate and 'live' set of information, will not only allow a more efficient use of assets, but will also speed up the retrieval of asset data, enabling more efficient execution of asset replacement and/or maintenance activities. The remote accessibility of the information will ensure that the Helpdesk and on-site operatives are more informed, and the use of BIM will instil a level of discipline on the data management of an asset very early on, making the ongoing life cycle management a more valuable exercise as well as creating a more accurate knowledge base for benchmarking.

Lessons learned

A key lesson learned during the sample school project was that it is a much bigger exercise to backfill data after the event rather than take the time to develop the required data structures in the first instance. The BIM model, though sophisticated, was itself restricted as it could not provide estimation software with the data requirements for life cycle costing. It required an external application to produce a complete life cycle cost analysis. There are, however, tools in the market that have demonstrated a generic template, which could be incorporated within the BIM process to address some of the barriers to successful implementation of WLC in practice. It is, therefore, not unreasonable to assume that further developments in the area of life cycle costing within the model environment will address these limitations.

Obtaining the right and relevant information from the supply chain also proved to be a big task, which took some time as it involved an iterative process with suppliers to encourage them to provide the correct data. This demonstrated that all relevant parties involved in the BIM process need to speak a common language, and have a common understanding of terminology and requisite data.

It is also important that all parties understand what systems are being used to undertake what functions, and that there is clarity on the information exchange required. Not all stakeholders in the process need the same skills, but all require a base level of understanding of terminology and roles and responsibilities.

Summary

This chapter sought to describe how BIM was applied in the area of whole life costing. The content of the case study was based on a 'sample' school project commissioned by an in-house multidisciplinary team and accountable for the development of a BIM capability and its implementation across a global engineering enterprise.

The chapter appraised the approach taken in the application of BIM, the processes and tools utilised, the challenges faced, and the opportunities for further development. The result was the achievement of an integrated and collaborative process between the team members. This in itself enabled a greater level of integration at every stage of delivery – from feasibility through construction and on to operation, and which should ultimately also be useful in the decommissioning phase.

BIM is emerging as an innovative way to manage projects. Asset performance and predictability of outcomes are greatly improved by adopting BIM. As the use of BIM accelerates, collaboration within project teams should increase, which will lead to improved profitability, reduced costs, better time management and improved customer/client relationships (Azhar et al., 2008). The pressure from clients and building or facility owners demanding more efficient, and economic assets, will inevitably lead them to utilise BIM on their projects.

There has been a language gap between designers and constructors focused on capital expenditure, and asset managers and property teams focused on operations. Whilst this gap is gradually closing, it still depicts a level of perception of the value life cycle costing brings to the wider built environment. At a strategic level, many design and construction teams still operate without reaching out to the asset managers and property teams, limiting the level of knowledge that can be unlocked to drive long-term value on projects.

On a technology level, there are still issues around data exchange and interoperability that need to be overcome. BIM requires the development of open systems and standardised data libraries of knowledge that can be utilised by downstream asset management systems. Without such standardisation, effective benchmarking and collaboration becomes very difficult and costly.

Barker (2014) in his report on 'digital engineering', opines that having the end in mind is a good starting point for the journey and developing a future process, from design intent to site installation and on to inspection and operation, will change the way projects are delivered and will give real benefits in terms of productivity.

BIM is being increasingly used to assist built environment professionals to conceptualise, design, construct, and ultimately operate built assets. The BIM environment provides the ideal platform to perform cost and value management exercises in order to achieve the most economical and sustainable building solution; and this is a key philosophy behind the drive of whole life costing.

References

Azhar, S., Hein, M., and Sketo, B. (2008) 'Building Information Modelling (BIM): benefits, risks and challenges' *Proceedings of the 44th ASC National Conference*, Auburn, AL, April 2–5.

Barker, D. (2014) 'EEJ Digital Engineering: "Moving to level three"'. Laing O'Rourke. Online. Available HTTP: <www.laingorourke.com/engineering-the-future/digital-engineering/eej/moving-to-level-three.aspx>, accessed April 2015.

BSI (2008) BS ISO 15686-5:2008 *Buildings and constructed assets: service life planning: Part 5, life-cycle costing.* British Standards Institute, London.

Godager, B. (2011) 'Analysis of the information needs for existing buildings for integration in modern BIM-based building information management', *Proceedings of the 8th International Conference*, May 19–20, 2011, Vilnius, Lithuania.

Hallberg, D. and Tarandi, V. (2011) 'On the use of open BIM and 4D visualisation in a predictive life cycle management system for construction works', *ITcon*, 16: 445.

Institute of Civil Engineering, ICE (2012), 'ICE IS Panel, Policy Position Statement', *BIM, Building Information Modelling and Management* – Version 5, ICE.

Kelly, J. and Hunter, K. (2009) *RICS Research Report : Life Cycle Costing of Sustainable Design*, RICS, London.

Obieli, C. (2011) 'Construction process engineer: a new breed of professionals', *Infoworks Magazine*, Autumn 2011, p.12.

Office of Government Commerce (2003) *Procurement Guide, 07, Whole-life costing and Cost Management*, OGC, London.

Pittard, S. (2011) 'BIM concept and impact: a new dimension'. RICS iSURV. Online. Available HTTP: <http://www.isurv.com/site/scripts/documents_info.aspx?documentID=6183>, accessed 1 December 2014.

Shen, Z. and Issa, R.R.A. (2010) 'Quantitative evaluation of the BIM-assisted construction detailed cost estimates', *Journal of Information Technology in Construction*, 15: 234–257.

Shen, Z., Issa, R.R.A. and Gu, L (2007) 'Semantic 3D CAD and its Applications in Construction Industry: an outlook of construction data visualization', *Advances in Visual Information Systems: Lecture Notes in Computer Science*, v. 4781, pp. 461–467.

Sylvester, K. and Dietrich, C. (2010) 'Evaluation of Building Information Modelling (BIM) estimating methods in construction education', *Proceedings of the 46th ASC Annual International Conference*, Wentworth Institute of Technology, Boston, Massachusetts, April 7–10. Online. Available HTTP: <http://ascpro.ascweb.org/chair/pa>, accessed April 2015.

"WHY ON EARTH DID WE ACCEPT THE LOWEST TENDER?"

5 Procurement

Adrian Malone

Introduction

It is essential to ensure that BIM requirements are embedded in the procurement of a project. Through procurement BIM deliverables are defined by the client, priced and accepted by suppliers. Procurement provides the opportunity to evaluate suppliers' capability and approach, testing that this approach is compatible with the client's requirements which are derived from their asset strategy. Procurement also provides an opportunity for suppliers to give feedback to the client about whether their aspirations for the project are realistic and achievable.

This chapter will explore the process and purpose of procurement for capital expenditure (CAPEX) projects where BIM Level 2 is applied to the project in accordance with PAS1192-2: 2013. It will examine the position of procurement within the overall asset and project lifecycle and the link between procurement and both the client asset strategy, and the important role procurement plays in establishing a connection between procurement for CAPEX and operation and maintenance of an asset post-handover (OPEX) through specification of information requirements and implementation of soft landings.

Author biography

Adrian Malone is currently Group Head of Knowledge Management and Collaboration at Atkins and was previously Head of BIM and Knowledge Management at Faithful+Gould, a division of the Atkins Group. He has 20 years' experience working in the construction industry. Adrian is a member of the RICS Global BIM Working Group, and is a RICS Certified BIM Manager.

Adrian is a founding committee member of the APM (Association for Project Management) Knowledge SIG (Specific Interest Group). He speaks regularly at conferences and industry events on BIM and collaborative working, and has written a regular blog column on BIM for *Building Magazine*.

Adrian has a Master's degree in Information Systems.

Company information

Faithful+Gould is a worldwide integrated project and programme management consultancy, supporting clients with the construction and management of their key assets, important projects and programmes. High-profile projects include the World Trade Center site and ITER. Clients include such companies as GSK, EDF, Michelin, Rolls Royce, Intel, Exxon and Coca-Cola Co.

With a turnover in excess of £230m, Faithful+Gould employ over 2,300 professionals operating from a network of more than 50 offices throughout Asia Pacific, UK and Europe, the Middle East and the Americas.

Faithful+Gould is a member of the Atkins Group, a global design, engineering and project management consultancy.

Background to case study featured

The case study provided in this chapter examines how BIM was integrated into the quantity surveyor procurement role across a series of projects at Birmingham City University (BCU). Birmingham City University is one of the UK's largest universities, serving around 25,000 students, and is situated in Britain's second largest city. As part of its legacy strategy the University made a £180m investment, helping to consolidate its assets and including the creation of two new buildings within the heart of the city centre campus, the Birmingham Institute of Art and Design (BIAD), and a new Student Centre.

The University is one of the early pioneers of BIM, recognising the value BIM could bring to both the management of the design and construction processes, but also BIM's value for the operational management and maintenance of assets post-handover. The University's adoption of BIM preceded the publication of the UK *Government Construction Strategy* in 2011 and was implemented before key documents such as PAS1192-2:2013 were created. Subsequently the University has commenced work to integrate the BIM process set out in PAS1192-2:2013, and it was one of the first clients in the UK to adopt the use of the Employer's Information Requirements (EIR) document into the tender process in 2014 for the development of a new Conservatoire in the heart of Birmingham.

The first building to be commissioned by the University utilising BIM, The Parkside Building, was handed over in June 2013. The building was handed over on time and within budget with a fully coordinated clash resolved 3D model delivered to the client. A three-year soft landings period was commissioned as part of the main contractor and their Mechanical and Electrical (M&E) subcontractor on appointment. Following handover BIM has been used to support the operational management of the facility, with a federated 3D model of the building available to the maintenance team on a tablet device.

Case Study 1: Phase 1 City Centre Campus, The Parkside Building

Service profile

The service provided was traditional quantity surveying advice to the client and the design team.

Scope of service

Faithful+Gould were contracted to provide QS and Employer's Agent services. Procurement services included:

- Assisting University with procurement of contractor via the OJEU process.
- Development of procurement strategy in conjunction with the client.
- Procurement workshops including objective evaluation of preferred procurement routes.

Case study details

Names of parties

- Client: Birmingham City University (BCU)
- Project Managers/Lead Consultant: Birmingham City University/Associated Architects

Figure 5.1 Completed Parkside Building (left) and Curzon Building – under construction (right).
Photo courtesy of A. Malone.

- Architect: Associated Architects
- Engineer: Ramboll
- QS: Faithful+Gould
- Contractor: Willmott Dixon
- Building Services Engineer: Arup.

Details of project

The new building houses Birmingham Institute of Art and Design (BIAD) as well as Media Production for the Faculty of Performance Media and English (PME). The Parkside Building provides a media hub offering courses in TV, radio and photography; generic studio space; wood/metal working workshops; ceramics and glass workshops; dye/printing and weaving studios; support functions; general teaching space and linkages to the School of Architecture and Building. The project provides a gross floor area of 18,300m².

Project value

The project value was £42m

Dates

August 2011 to June 2013

How BIM was used

The use of BIM for The Parkside Building was initiated before the publication of the UK Government Construction Strategy and so members of the supply chain team were at or near the beginning of their BIM journeys. Guidance such as PAS1192-2:2013 and the digital plan of work had not been created, and there were few, if any, relevant case studies about the use of BIM on similar projects to draw upon. The consultant team for The Parkside Building was appointed by the client prior to BIM being identified as a requirement. During the early concept stage of the project the development site was moved from its former location to a nearby site adjoining Millennium Point due to the original site being acquired by HS2 (High Speed 2) for its new Birmingham Terminal. It was during this time that the requirement for BIM was introduced by the client, BIM was therefore not a consideration in the appointment of the consultant team. When BIM was raised by the client this was met with some initial trepidation by some members of the consultant team, but the use of BIM was bought into by all. It was clear from the start that the client wanted a clash model which would be used to operate and maintain the building post-handover.

At concept stage the design was represented through photographs of hand-crafted models created from plasticine, subsequently a software drawing tool was used by the architect to create the initial 3D models. Faithful+Gould looked at the viability of using the drawing tool models for quantity take-off, but at the time the cost estimating software used supported only DWFX files for transfer of model data, and the drawing tool could not provide compatible information. For The Parkside Building, BIM was therefore not used at the early concept stage, rather a traditional cost model was created. A basic model in BIM would have been useful at this stage to allow rapid quantity take-off of volumes and to explore the cost impact of different design options.

Birmingham City University retained the project management role in-house and led the contractor procurement process, with the QS supporting tender and procurement processes. The use of BIM was an explicit requirement for the appointment of the contractor, who was required to produce a fully coordinated, clash free BIM model. Subsequently the term 'clash resolved' was substituted for 'clash free' as experience demonstrated that it was not cost effective or necessary to remove all clashes from the model, as those which were consequences of modelling methodology could simply be accepted.

Prior to the publication of PAS1192-2:2013 the term Employer's Information Requirements (EIR) was not in currency. An Employer's Requirements document was prepared by the client with input from the consultant team. This document established the requirement for a fully clash resolved BIM model with associated documentation to be created during the design and construction phase as a deliverable, which was to be passed to the client upon handover to support the operation and maintenance of the asset in use. This task was assisted by the appointment of a BIM Process Manager (BPM) within the University Estates team. This appointment was made during the design and construction phase of The Parkside Building, and the BPM worked closely with the consultant team and contractor to ensure that operation and maintenance information requirements were built into the model during construction.

A two-stage competitive tender process was adopted to select a contractor under a Design and Build contract. Within the tender process contractors bidding for the scheme were required to develop a BIM Execution Plan; a fully coordinated clash resolved BIM was a key requirement upon completion.

Following early concept stage a coordinated BIM model had been developed by the architect, structural engineer and MEP consultants up to stage D (under the RIBA Plan of Work 2007). This model was used for cost estimation by the quantity surveyor. It is interesting to note the approach which was taken to adoption of BIM here. There were very few examples of BIM in the UK at the time, and protocols were insufficiently mature to support the exchange and coordination of information via BIM. The model was exported into DWFX format so that it could be read by the software in use by the quantity surveyor. Reliance and trust in the model for quantification was an iterative process. Initially 2D CAD drawings were measured electronically for quantity take-off, quantification from the 3D model was then undertaken and compared with the 2D measurement. As confidence and experience grew the process was reversed, with digital quantification from the 3D model undertaken first, and spot check measurements from the 2D CAD files used with decreasing frequency.

The Stage D model was issued to the contractors as part of the tender pack. Development of the BIM model did not form part of the tender requirements, for the contractors however, BIM was used by the QS to validate contractor submissions with quantities extracted in order to validate tender submissions for completeness and correctness. BIM was a valuable tool in this process, allowing rapid checking of contractor's proposals against the Stage D design which had been issued at tender.

Evaluation of contractor's submissions included an evaluation of the BIM Execution Plans submitted. This evaluation was undertaken by the client and the consultant team. This evaluation was important as it was recognised that development of the digital output alongside the physical building would require the right focus on information coordination and process.

Case Study 2: Phase 2 City Centre Campus, Curzon Building

Service profile

The service provided was traditional quantity surveying advice to the client and the design team.

Scope of service

Procurement services included:

- Assisting University with procurement of contractor via the OJEU process.
- Development of procurement strategy in conjunction with the client.
- Procurement workshops including objective evaluation of preferred procurement routes.

Figure 5.2 Site works in progress for Curzon Building in foreground, and Parkside Building in background.
Image copyright M. Hamilton-Knight.

Case study details

Names of parties:

- Client: Birmingham City University (BCU)
- Project Managers Lead Consultant: Birmingham City University (PM)/ Associated Architects (LC)
- Architect: Associated Architects
- Engineer: White Young Green (WYG)
- QS: Faithful+Gould
- Contractor: Willmott Dixon
- Building Services Engineer: Hoare Lea

Details of project

The Curzon Building creates a new student centre with functions that include a Business School; Faculty of Law and Social Sciences, integrated food and catering offer; Library; Student Centre; Social Learning Space; General teaching space; and lecture theatres. Starting in June 2013 a new facility covering 23,200m² is being created with a construction value of £46m. Faithful+Gould were contracted to provide QS, CDM-C, Party Wall and Right to Light services.

Project value

The project value was £46m

Dates

June 2013 to May 2015

How BIM was used

As with The Parkside Building, Faithful+Gould were appointed quantity surveyor. The Student Centre was less complex than The Parkside Building which contained a number of specialist areas such as film and photography studies, but it was equally important to the University and students. The architect employed for The Parkside Building was appointed to the Student Centre project which provided some continuity to the BIM process, with different appointments made for MEP and structural consultants.

Lessons learnt from the adoption of BIM for The Parkside Building were brought forward into the project for the Student Centre. Once again a primary driver for the use of BIM by BCU was to deliver a fully coordinated clash resolved 3D BIM model which was defined within the project requirements for use in the management of the operation and maintenance of the facility post-handover. The Employer's Requirements document provided for The Parkside Building was updated by the client with lessons learnt from that project. As noted in Case Study 1, this task was assisted by the appointment of a BIM Process Manager (BPM) within the University Estates team to work closely with the consultant team and contractor to ensure that operation and maintenance information requirements were built into the model during construction. The Student Centre project was onsite at the time of handover of The Parkside Building, but important knowledge and lessons had been learnt from the first BIM project. The BPM provided continuity for the University and was able to help shape the Employer's Requirements for the Student Centre with knowledge gained from the development of The Parkside Building.

Early conceptual design was undertaken by the architect to develop a coordinated and federated BIM model containing architectural, structural and MEP design information. Once again the quantity surveyor was able to make use of the BIM model, exported to DWFX file format to take off quantities directly from the model in order to provide cost estimates.

Procurement of a contractor for the Student Centre was initially explored via an existing framework, with design developed to Stage E (under the RIBA Plan of Work 2007). Subsequently, a decision was made to use a two-stage competitive tender process under a Design and Build contract. As with The Parkside Building, the Employer's Requirements document and federated BIM model were issued as part of the tender pack.

Case Study 3: Birmingham Conservatoire

Service profile

The service provided was traditional quantity surveying advice to the client and the design team.

Scope of service

Faithful+Gould were contracted to provide QS and CDM-C services. Procurement services included:

- Assisting University with procurement of contractor via the OJEU process.
- Development of procurement strategy in conjunction with the client.
- Procurement workshops including objective evaluation of preferred procurement routes.

Case study details

Names of parties:

- Client: Birmingham City University
- Project Managers/Lead Consultant: Birmingham City University (PM), Fielden Clegg Bradley Studios (LC)

- Architect: Fielden Clegg Bradley Studios
- Structural Engineer: WYG
- Building Services Engineer: Hoare Lea
- QS: Faithful+Gould
- Contractor: TBC

Details of project

The third and most recent project in the case study has been included to show how the use of BIM at BCU continues to develop as industry practice and published standards develop. The Birmingham Conservatoire will create a new Conservatoire space in the heart of Birmingham. The current Birmingham Conservatoire, which was originally founded in 1886, provides training for orchestral, jazz and opera performers as well as conducting research into composition, musicology and performance with live electronics. The existing building will be demolished as part of the £450m redevelopment of Paradise Circus.

The new unique, contemporary-style building will provide state-of-the-art facilities for students. It will incorporate two major performance spaces including a new concert hall for orchestral training and public performance, private rehearsal and practice rooms, recording studios and new technology, in addition to providing teaching spaces for musicians from a variety of disciplines.

Project value

Project value circa £27m–32m

Dates

August 2015 to August 2017 (estimated at time of publication)

How BIM was used

Birmingham City University continued to evolve its use of BIM and to stay ahead of the curve as industry maturity develops. The Parkside Building and Student Centre were both initiated prior to the issue of PAS1192-2:2013 and associated development of supporting guidance and protocol documents. The development of a new Conservatoire by the University to replace the existing Birmingham Conservatoire which would be demolished as part of wider regeneration in the city provided an opportunity to update the University's processes to reflect practice established in PAS1192-2:2013 published in February 2013, the tender process to appoint the consultant team was opened in January 2014.

The University adopted the use of the Employer's Information Requirements (EIR) document to establish client BIM requirements within the tender process. Suppliers were required to respond with a BIM Execution Plan detailing capability, competency and approach to realising the requirements set out in the EIR as

Table 5.1 Core content of the EIR.

Technical	Management	Commercial
Software platform	Standards	Data Drops and Project
Data Exchange	Roles and Responsibilities	Deliverables
Format	Security	Clients Strategic Purpose
Co-ordinates	Co-ordination and clash	BIM Specific Competence
Level of Detail	avoidance	Assessment
Training	Collaboration process	
	Health and safety / Construction	
	Design Management	
	Systems Performance	
	Compliance Plan	
	Delivery Strategy for Asset	
	Information	

part of their tender submission. Faithful+Gould were successful and as such were appointed in the role of quantity surveyor and CDM Coordinator (CDM-C). It was anticipated that the consultant team would prepare a combined post-contract BEP, and that the EIR would be used within the tender process of the contractor.

The EIR was prepared by the University under the three core headings: Technical, Management and Commercial (see Table 5.1).

Why was BIM used and who made the decision?

The decision to use BIM was made by the University Estates team who saw BIM as a natural progression to their use of structured databases of asset information used to manage the University estate. With responsibility for the operational management and maintenance of the estate, the team's objective was to create a single database which combined both a 3D graphical model of their assets and associated data and related documentation necessary for operation and maintenance. Benefits during design and construction were also sought, with BIM used extensively to support stakeholder engagement and design coordination during design, and the use of BIM integrated with programming of construction works tested during construction. Faithful+Gould worked with BCU and the wider project teams to explore the use of BIM for cost analysis and in support of procurement processes.

Procurement under BIM

The procurement process is the most important opportunity to successfully define client requirements. Correctly identifying and articulating requirements at this stage increases the likelihood that specific project requirements will be understood, resourced, costed and ultimately delivered by the successful project team. Procurement is also an opportunity to test all requirements in order to weed out any which could add to the cost of the project without delivering added value to the client.

It has always been a function of the procurement process to identify client project outcomes and to ensure that these are reflected adequately both in the selection of procurement route, and in the specification of tender requirements. When procuring a project in which BIM will be adopted all of the considerations which would be required in a traditional project remain valid. BIM introduces some additional considerations, most significantly where asset information to be used during maintenance and operation is specified during the CAPEX procurement. These additional requirements must be carefully specified to ensure that the Asset Information Model (AIM) or Construction Operations Building Information Exchange (COBie) file contain information coded and structured and at a level of detail sufficient to support OPEX requirements.

Soft Landings

BIM sits within a wider construction strategy, and a wider approach to project delivery. Soft landings, first defined by BSRIA, have been adopted by the UK Government to form Government Soft Landings (GSL).

The focus of soft landings is the handover of a built asset from construction into operational use (Figure 5.3). The purpose of a soft landings policy is to ensure that necessary arrangements are in place to increase the likelihood that handover will be a smooth and effective process. These arrangements may include contractual requirements which specify that members of the construction team remain available and responsible to support operational use in the first few years (typically 3 years) post-handover. Soft landings may also require

Figure 5.3 Gap between delivery and handover.

a support package to ensure that end users and maintenance teams are equipped with the information and understanding of the asset enough to effectively operate it from day one.

Whilst a soft landings framework is focused on the handover stage, its definition and requirements must be considered from the outset and reflected in the procurement strategy and in the requirements set out in the tender process.

Procurement of digital asset

Under a traditional (non-BIM) construction project the primary deliverable would be the physical asset – i.e. a new building or infrastructure asset, or the refurbishment of an existing asset. When implementing BIM, a digital asset may be procured alongside the physical asset. A digital asset typically means a computer model of the completed building(s) or asset(s) which represents what has been constructed.

Level 2 BIM in the public sector defines the required digital asset as a COBie file. A COBIE file is a structured spreadsheet document containing all specified information about the asset – including basic geometry alongside specification of building elements.

In the private sector, COBie is not a mandatory requirement, but may be required by some clients. Other clients may require an as-built model of the asset and handover in a format other than COBIE, such as IFC (an interoperable file format associated with BIM, defined by BuildingSMART), or in a proprietary software format for example in order to incorporate into the client's CAFM (Computer-Aided Facilities Management) system.

PAS1192-2:2013, published by BSI, was specified as a requirement for BIM Level 2 by the UK Government BIM Task Group in 2014. BIM Level 2 was mandated for all centrally procured UK construction projects by 2016. Whilst mandated only for centrally procured government projects, many local government clients and clients in the private sector are specifying BIM Level 2 and/or adoption of PAS1192-2:2013 as a client requirement through their procurement process.

Key stages of PAS1192-2:2013 for procurement

PAS1192-2:2013 and PAS1192-3:2014 together define the information process cycle around the lifecycle of a built asset. PAS1192-2:2013 focuses on CAPEX, with PAS1192-3:2014 setting out requirements for OPEX. PAS1192-2:2013 and PAS1192-3:2014 has been designed to be agnostic to procurement route and contract type, it may be implemented with any of the contract types which were in existence at the time of its publication.

PAS1192-2:2013 sets out a process for procurement of capital works. The process steps outlined may be applied once, or as an iterative process – for example the process may be used to bring on board the consultant team for a project, and then subsequently to appoint the contractor. The contractor may then apply PAS1192-2:2013 to the procurement of sub-contractors and so on.

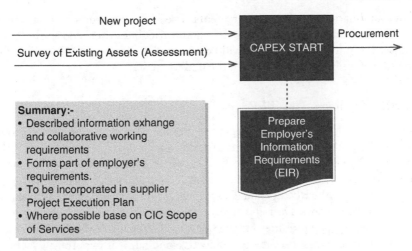

Figure 5.4 Preparation of the EIR.

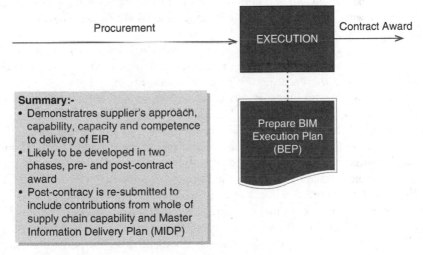

Figure 5.5 Preparation of the pre-contract BEP.

The Employer's Information Requirements (EIRs)

The BIM process as defined by PAS1192-2:2013 commences when a requirement for capital expenditure is identified. This requirement may relate to the creation of a new asset, or a requirement for the refurbishment, demolition or alteration of an existing built asset.

The EIR documents commercial, management and technical client requirements in relation to the use of BIM for a project. The EIR is prepared by the client, possibly with support from a BIM consultant or an externally appointed project manager employed by the client (Figure 5.4). The EIR sets out the client's strategic purpose, project outcomes, project-specific BIM requirements and the client's information exchange requirements. The EIR is important as it is through the creation of this document that the client articulates their end goal – what the asset is for, along with how it will be operated and managed. Crucial to the formation of an EIR is to 'begin with the end in mind'.

The BIM Execution Plan (BEP)

For procurement, the EIR forms part of the procurement pack. Suppliers respond to the EIR through the procurement process by submitting a pre-contract BIM Execution Plan (BEP). Within the BEP suppliers demonstrate that they have understood the client's requirements which have been set out in the EIR, and they demonstrate their competence, capacity, capability and approach to meeting those requirements. During a typical process to procure the main contractor, there will be opportunities for the quantity surveyor to lead the evaluation process. To ensure that assessment is objective and fair it is important to set out clearly how BIM responses will be evaluated when the procurement strategy is defined. This will enable appropriate provision to be made in the procurement to comply with any relevant rules and requirements governing assessment and award.

Depending on the procurement strategy, the PAS1192-2:2013 process will be applied as each member of the project team is appointed. For example the design team and the construction team are often appointed through separate and successive tender processes. In instances where appointments are undertaken in stages, the PAS1192-2:2013 process can be applied to each successive supplier appointment. Under this arrangement the quantity surveyor and other consultants should have in a previous iteration of the PAS1192-2:2013 process submitted their own BEP. The consultant team may bring together their separate pre-contract BEPs into a single post-contract BEP (Figure 5.5). The BEP should be viewed as a live document which continues to evolve and reflect additional supplier appointments and requirements as the project develops. The post-contract BEP will, therefore, evolve as appointments are made during the delivery phase of the project – for example upon appointment of the contactor.

New procurement models

The CIC BIM Protocol, published by the Construction Industry Council (CIC) was developed to enable Level 2 BIM to be implemented through any of the forms of procurement prevalent in the UK construction industry at the time, and it forms one of the seven component parts of Level 2 BIM as defined by the BIM Task Group. The CIC BIM Protocol provides flexibility, removing compatibility with procurement as a barrier for the uptake of BIM. This is explained in detail in Chapter 7. It is widely recognised in the construction industry that whilst the CIC BIM Protocol retains flexibility for selection of procurement route, the choice of route can have significant impact on the implementation of BIM in practice.

The Procurement/Lean Client Task Group was established by the Government in 2011. One key area of focus for the Group was to examine procurement models which would best support the adoption of BIM and the wider UK Government Construction Strategy. Three procurement models were explored in detail, each supporting early contractor involvement and collaboration, both of which are beneficial when working with BIM. Three trials were established, each focused on a different procurement model, these being: Cost-led procurement utilising NEC 3 Option C, Integrated Project Insurance (IPI) using PPC 2000, and Two Stage Open Book utilising JCT Constructing Excellence. Trial projects were undertaken within the UK Government BIM programme to evaluate the use of each of these approaches in practice. In 2014 detailed Guidance was published by the UK Government setting out information to support the use of each procurement model in detail and publishing benchmark data where available from the trial projects. References for each of these three procurement models are provided below.

Two Stage Open Book

'Two Stage Open Book' is a system of preconstruction phase project processes governed by the early appointment of a full project team (Cabinet Office, 2014c).

In this method of procurement the different team members (architects, QSs, Tier 1 contractors, etc.) are asked to tender for the project on the basis of an outline brief and a cost benchmark. This selection is based on capacity, capability, plus their profit/fees/overheads etc. From this Stage 1 competition the successful participants are appointed to work up the detailed proposals on the basis of an open book cost, which meets the client's requirements and the cost benchmark as the second stage of the process.

This method allows the client to select the individual team members at Stage 1 and then encourages all the team members to work together collaboratively during Stage 2 to deliver the client's requirement. Successful completion of Stage 2 then leads onto the delivery of the project by the team that has built successful relationships during the second stage of the procurement process.

Further guidance on this procurement method can be found at:

https://www.gov.uk/government/uploads/system/uploads/attachment_data/file/325014/Two_Stage_Open_Book_Guidance.pdf

Cost-Led Procurement

The client provides a strategic brief outlining what is required, by when, and to what standards etc. The client then engages with the suppliers who form project-specific teams to respond to the strategic brief. The teams will typically comprise the design consultants, Tier 1 contractors and key sub-contractors. The initial competition takes place as early in the project life as is practical, with the teams committing to deliver the project below a cost ceiling which has been set. This is the key feature of this procurement method.

A two-stage process is also adopted in this method, with the potential to take two teams through to a second stage, to refine and develop their proposals.

It is important that the successful team demonstrates the ability to meet and better the cost ceiling at inception and achieve this as an out-turn cost. The team offering the best solution and cost is appointed and requested to work with key client stakeholders to develop the design and cost in parallel.

This method of procurement, while focusing on the cost ceiling as with Two Stage Open Book, enables the client to see the evidence of the different teams working together and collaborating to deliver the detailed proposal which can then move into delivery.

Further guidance on this procurement method can be found at:

> https://www.gov.uk/government/uploads/system/uploads/attachment_data/
> file/325012/Cost_Led_Procurement_Guidance.pdf

Integrated Project Insurance

As the name implies this method is centred round an Insurance Policy. As with the other methods the client develops a strategic brief outlining what is required, by when and setting standards, etc.

This method does not prescribe how the suppliers are selected; it is, however, likely that teams will form to respond to the strategic brief. The successful team will then either form an Alliance or a virtual company to deliver the client's requirements. The key criteria for the method is that there is a no blame/no claim undertaking between the Alliance members, and the whole project is underpinned by an Integrated Project Insurance (IPI) policy which insures any cost overruns above a pain share limit.

The premise for this method is that IPI should cost no more than the traditional project insurances, but the no blame/no claim culture which is engendered will bring with it the real benefits of collaboration as it is in all parties' interests to deliver a successful project and to reduce waste and improved performance as all will benefit. There is also positive encouragement to surface all problems or issues as soon as they become evident, to deliver the best solutions as quickly as possible.

Further guidance on this method of procurement can be found at:

> https://www.gov.uk/government/uploads/system/uploads/attachment_data/
> file/326716/20140702_IPI_Guidance_3_July_2014.pdf

Issues/benefits

Benefits

The case study describes how the use of BIM at Birmingham City University evolved between 2008/9 and 2014. The use of BIM for The Parkside Building was truly pioneering, and in the absence of guidance subsequently published (e.g.

PAS1192-2:2013) or comparable BIM projects to learn from the University and the project team were required to design a process which would allow them to work together to effectively, deliver a successful construction project, whilst creating a fully coordinated clash resolved model in BIM which is now in use by the client to operate and maintain as part of their estate. The Student Centre project commenced mid-way through the delivery of The Parkside Building, however, sufficient time had elapsed to allow lessons and experience gained from The Parkside Building to be fed into the procurement and subsequent design and delivery of the Student Centre project. Standards and guidance have caught up with the publication of PAS1192-2:2013 and the University has continued to evolve its BIM process, adopting the Employer's Information Requirements (EIR) document in place of the Employer's Requirements document used on the previous two schemes, and adopting the use of the BIM Execution Plan (BEP) as a requirement for tender responses and to facilitate the coordinated working of the project team.

The use of BIM brings some significant benefits and opportunities to the quantity surveyor in supporting the client and consultant team through procurement of the main contractor. Across the projects described in the case study Faithful+Gould have found that BIM supports the communication of the design more effectively than 2D CAD drawings, this allows the quantity surveyor to gain a richer understanding of the design from the model files and consequently the need to query design intent is reduced.

BIM helps to ensure that all design information is communicated. It is not unusual for traditional design information packs to be missing particular schedules, elevations, etc. which are required by the QS, typically this causes delay as the QS requests this information from the design team. Through the use of BIM the QS is able to navigate through the model, and where necessary Faithful+Gould have been able to extract sections and elevations useful to undertake analysis for quantification purposes but which were not provided within the document pack provided by the design team which are needed for analysis. Currently this requires working directly with design software, although, increasingly, such functionality is being added to measurement and estimating software.

A key benefit to the QS and the wider project team is speed. Quantification from BIM is significantly more efficient than from 2D CAD where manual measurement is necessary. Through a vastly more efficient quantification process the quantity surveyor is able to give more responsive feedback about different design options.

Challenges

BIM brings many benefits to the quantity surveyor during procurement, however, challenges remain. The estimating software used by the quantity surveyor for the Birmingham City University schemes was not able to open native BIM authoring tool files, resulting in the use of a reduced export format DWFX. The QS team found that DWFX did not in all instances provide the level of detail needed. Faithful+Gould resolved this by using a BIM authoring tool where necessary to interrogate the model in its native file format. Estimating software is increasingly providing capability to read BIM authored files directly, and compatibility and consistent implementation of the Industry Foundation Classes (IFC) schema has improved considerably since 2008/9 when the use of BIM at BCU was initiated, and is, therefore, unlikely to be a long-term issue.

A greater challenge derives from the differing needs of the quantity surveyor and the design team in terms of model structure and the way in which information is organised. BIM provides a process through which information exchange requirements can be established and embedded into project protocols within the BIM Execution Plan, it should be noted however that it is not always feasible for the design team's modelling methodology to fully meet the requirements of the quantity surveyor's team for information extraction, for example temporary works may not be fully specified in the model, yet the quantity surveyor must account for these in the cost estimate. Dialogue between the QS and the wider consultant team is essential from the earliest opportunity to create a shared understanding of what is needed by each party, and what can be achieved within the constraints of programme and budget, as well as both supplier and software capability. If the quantity surveyor is appointed in parallel with the design team there is no opportunity for the quantity surveyor's requirements to be specified in procurement (or indeed for those of the other members of the consultant team), early discussion between the appointed consultants is therefore essential to establish mutually beneficial working practices. Industry standards can help establish collaborative working protocols between suppliers; professional institutions and industry bodies therefore have an important role to play in establishing working practices at an industry level which can then be applied effectively at the project or portfolio level.

Key benefits resulting from the use of BIM

To the project

The key benefit achieved through the use of BIM was the successful creation of a fully coordinated clash resolved 3D BIM model which is successfully being used by Birmingham City University for the operation and maintenance of The Parkside Building post-handover. Whilst there is much room for improvement and the University continues to evolve its use of BIM, the provision of a digital model at handover has been sufficiently valuable for the University to continue to specify and develop this requirement through subsequent projects and to begin considering an approach to the digitisation of the whole estate.

To the business

The benefits realised by Faithful+Gould through the use of BIM at Birmingham City University and on other projects have been both technical and non-technical. At a technical process level BIM allows for more efficient and accurate quantity take-off which supports the provision of timely and accurate professional advice to our clients. BIM means more efficient quantification, extracting such data from the model is more efficient than manual measurement of dimensions in CAD, BIM also encourages earlier dialogue between the QS team and the wider project team, which can provide opportunity for more effective collaboration on the project, which in turn supports better results for the client. BIM also better communicates the design when compared to 2D CAD resulting in a reduced need to seek clarification from the design team.

To others

The use of BIM for the development of The Parkside Building supported rich engagement with stakeholders, such as end users of the building during design stage. This has helped to ensure that the building upon completion meets the expectations and needs of the end users. In the future there is an opportunity to use BIM to support post-occupancy evaluation (POE) both to validate that expected performance in terms of energy efficiency specified at design stage have been realised, and also to support capture of feedback from end users and other stakeholders about how well the spaces created support their intended use.

Summary/commentary

The use of BIM requires suppliers working within a multidisciplinary project team to coordinate the way in which information contained within and associated with the model is created and managed in order that the client's outcomes for the use of BIM are met. A client's BIM outcomes may extend beyond the coordination of information within the capital project itself (CAPEX); for Birmingham City University outcomes relating to the use of BIM post-handover (OPEX) were the primary driver.

Procurement can play an important role in communicating the client's outcomes to potential suppliers, and establishing the principles (such as project team collaboration) which the client seeks to apply. The Employer's Information Requirements (EIR) document is a template which can be used to communicate client outcomes and requirements at an early stage as part of the procurement pack. The pre-contract BIM Execution Plan provides a common approach by which potential suppliers can demonstrate their capability, capacity, competence and approach to meeting the requirements defined by the client in their EIR, thereby allowing different supplier bids to be assessed and evaluated by the client and their consultants.

In practice it is not always possible to fully define requirements during initial procurement. As the project progresses further requirements may be identified, and depending on procurement model further appointments may be required after initial procurement has taken place. Such circumstances are common, and it is important to understand that the PAS1192:2-2:2013 process is not applied as a single loop tracking the project lifecycle, rather it may be applied iteratively to each successive round of procurement activity. The EIR and in particular the post-contract BEP should be refreshed as the project progresses and as further appointments are made through procurement.

The bridge between CAPEX and OPEX is an important consideration during procurement. BIM Level 2 specifies a soft landings process by which the requirements (both in terms of information and specification) of end users and those who will operate and maintain the asset post-handover should be incorporated early in the CAPEX process, it is essential that appropriate consideration is given to a soft landings as part of the procurement of a CAPEX project in order to comply with BIM Level 2 requirements but also as part of adopting a best practice approach.

Procurement can set the conditions for the successful implementation of BIM for capital projects, and through 'thinking with the end in mind' can help ensure that when an asset is handed over into operational use that it meets the needs and aspirations of end users and the requirements of facilities managers. Procurement lays the foundations upon which a successful BIM project is built.

It is perhaps worth noting that none of the case studies featured in this chapter make reference to the use of either e-tendering or e-procurement, and that, whilst neither are a prerequisite for BIM, they do offer the potential of a more holistic solution to support the project information lifecycle.

References

BSI (2013) *PAS1192-2 Specification for information management for the capital/delivery phase of construction projects using building information modelling*, British Standards Institute, London.

BSI (2014) PAS1192-3:2014 Specification for information management for the operational phase of assets using building information modelling, British Standards Institute, London

BSRIA BG 4 (2009) *The Soft Landings Framework, Report BG 4/2009*, Building Services Research and Information Association, Bracknell, UK.

Cabinet Office (2011) 'Government Construction Strategy'. Online. Available at <http://www.cabinetoffice.gov.uk/resource-library/government-construction-strategy>.

Cabinet Office (2014a) 'Cost led procurement guidance: guidance for the procurement and management of capital projects'. Online. Available at <https://www.gov.uk/government/uploads/system/uploads/attachment_data/file/325012/Cost_Led_Procurement_Guidance.pdf>.

Cabinet Office (2014b) 'The Integrated Project Insurance (IPI) Model: project procurement and delivery guidance'. Available at <https://www.gov.uk/government/uploads/system/uploads/attachment_data/file/326716/20140702_IPI_Guidance_3_July_2014.pdf>.

Cabinet Office (2014c) 'Project procurement and delivery guidance using Two Stage Open Book and Supply Chain Collaboration'. HMSO, London. Online. Available at <https://www.gov.uk/government/uploads/system/uploads/attachment_data/ file/325014/Two_Stage_Open_Book_Guidance.pdf>.

RIBA (2007) *Outline Plan of Work*. RIBA, London.

"I STILL CAN'T FIND THE DOCUMENT YOU WANTED....."

6   Information Management

Malcolm Taylor, Peter Sell and Tahir Ahmad

Introduction

This chapter illustrates the importance of Project Information Management (PIM) within a BIM environment.

It describes what BIM means to a major infrastructure project, the Crossrail programme in London. The case study looks at the journey undertaken by multiple disciplines and contracts to build the railway and to provide asset information for the operator and maintainer, whilst aligning to requirements set out by the UK government's BIM strategy.

Furthermore, it outlines how the organisation has adapted and refined its approach in the light of the strategy and outlines lessons learnt to date for wider use by the industry.

Author biographies

Malcolm Taylor – Head of Technical Information, Crossrail Ltd.

Malcolm is responsible for the BIM strategy and implementation; Asset Information Management; technical data management; document control; GIS; and configuration management. He is a Chartered Engineer – prior to joining Crossrail he was Rail Director for a major global consultancy working on many UK and international rail projects. Starting as a Graduate with London Underground, Malcolm has over 35 years' experience in the design, construction and maintenance of large-scale transportation projects around the world, with a particular emphasis on railway design, programme and project management.

Tahir Ahmad – Specialist BIM Coordinator, Crossrail Ltd.

Tahir Ahmad is the BIM Coordination Specialist for Crossrail in London. His current responsibilities include BIM performance analysis and improvement and management of the Crossrail BIM Information Academy: Enabling BIM strategies to work in practice. Tahir is utilising his Information Management and business analysis background to launch and lead initiatives to champion

BIM for Crossrail, extolling the efficiencies and outcomes integrated Engineering Information management can create. Thus enabling Crossrail to leverage the value of interoperable and accessible data sets at the heart of its operations to allow informed, smart infrastructure management at every stage of the project from design to operations.

Project information

Crossrail is among the most significant infrastructure projects ever undertaken in the UK. From improving journey times across London, to easing congestion and offering better connections, Crossrail will change the way people travel around the capital. The Crossrail programme comprises nine new stations and 42km of tunnels constructed beneath the busy streets of London, together with the associated depots and maintenance facilities (see Figure 6.1). With a £14.8bn project budget running from 2001–2019, the Crossrail programme will increase London's rail transport capacity by 10 per cent, and secure an estimated £42bn net benefit to the UK economy.

To provide the new railway, Crossrail Ltd (CRL) was created as a 50/50 joint venture company owned by Transport for London (TfL) and the Department for Transport (DfT). In December 2008 Crossrail Ltd became a fully owned subsidiary of TfL. As such it is a Non Departmental Public Body (NDPB) whose sole purpose is to deliver the Crossrail programme. With over 25 main design contracts and 83 construction and logistics contracts (often with multiple interfaces), Crossrail provides an example of a highly complex programme of projects where the potential for miscommunication, duplication, use of out-of-date documentation and data interface problems is significant. The complexity is caused not just by the scale of the project, but by the diversity

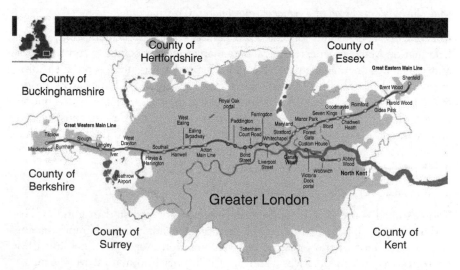

Figure 6.1 Crossrail location.

and interdependency of activities and disciplines both at station locations and along the route.

Most railway projects move into their operational phase with one maintainer. However, Crossrail will be operated by Rail for London (RfL) and have infrastructure maintained by Network Rail, RfL and London Underground Limited (LUL).

Crossrail had the objective to set world-class standards in creating and managing data, and the benefits of doing this have been significant, providing a single reliable source of information, reduced wastage, reduced cost and improved decision-making.

Service profile

Creating and bringing into use a railway as complex as Crossrail requires the highest standards of leadership, procurement, design and construction. The fundamental elements of any project are the control and balance of scope, cost and time. These variables, together with other underlying factors such as quality, must be mutually consistent and attainable. Information is the basic enabler behind all of these activities (see Figure 6.2). Data and information in a project – however big or small – is only created for one of two reasons:

1. it is legally required (by contracts; legislation; etc.)
2. to make decisions (e.g. schedules for managing; surveys for design; health and safety (H&S) records; test data for construction; asset attributes for maintenance; etc.)

Project Management is about bringing specific focus to the second of these two purposes. The more accurate, timely and relevant the data, the more efficient and effective the project delivery will be throughout the entire asset lifecycle (i.e. concept; design; construct; commission; operate and maintain; and decommission). PIM is an umbrella term that covers all the general systems and processes within

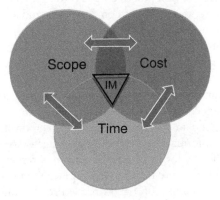

Figure 6.2 Information at the heart of project management.

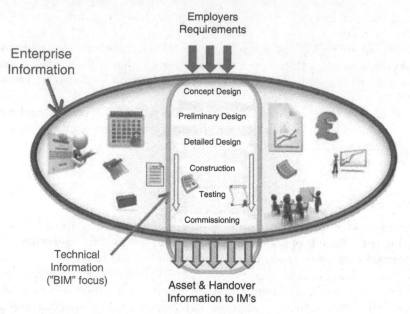

Figure 6.3 The scope of Project Information Management.

a project for the creation and use of information. This service is a subset of the overall Project Management responsibility.

Information Management (IM) within a project, and particularly a programme of projects like Crossrail, therefore, covers enterprise information for designing and building the railway, together with collecting and handing over the information for future operation and maintenance of the infrastructure and facilities (see Figure 6.3). Based on the experience of Crossrail, the authors see the role being based on a number of specific principles:

- Recognise the needs of a project: projects have a limited duration. It is, there-fore, important to use best practice from other projects, i.e. not to customise or modify software, which might compromise standards and consistency.
- Establish the information principles: only collect the information required and establish how this is to be done. For Crossrail, this has centred around the BIM approach of a collaborative common data environment (CDE) and setting out the required standards.
- Focus on benefits, outcomes and end-user needs of the information: from designers through contractors and onto maintainers, it is important to make sure that the **right** data is in the **right** place. In an infrastructure project, like Crossrail, some information is required specifically for building the railway, and some specifically for its operation and maintenance – appropriate data organisation and structure needs to be in place to ensure this happens.
- Prioritise project information needs: recognise information can have differ-ent 'sell-by' dates and risk profiles.

Table 2 of PAS1192-2:2013 provides some further guidance on the role of Information Management and specifically associated information exchange activities.

Treat information as a valuable resource – **it is** a valuable resource, just like people, plant and material, and needs to be treated as such. Getting the right behaviours in the use of information needs planning and continuous communications (in a similar way to health and safety).

In line with the principles outlined above, the PIM team is responsible for the following:

i.   Common data environment (CDE) – the creation, management and validation of the chosen CDE, involves working with the client and the rest of project team to determine the most appropriate CDE, instigating the environment and then managing the CDE through the life of the project.

ii. Ensuring the data within the CDE is appropriately structured so that it can be:

   a.   measured in terms of its data quality by appropriate metadata (similar to quality control of physical assets) for progress monitoring and key performance indicator (KPI) management

   b.   validated against the future maintainers' requirements and readily assured, to ensure data is acceptable for future maintenance activity information standards. The setting of the information standards to be adopted for the project will entail agreeing the standards with the client and the project team (see Figure 6.4 below).

iii. Ensuring that the information created matches the requirements for operation and maintenance in form and suitability, so that whatever system is ultimately chosen (e.g. for asset management), the appropriate data set can be made available.

iv. Information security – agreeing with the client and the project team the appropriate levels of information security (PAS1192-5:2015), including the information security marking of information to conform to the relevant standards.

v.   Master Data Management (MDM) – the creation, standardisation and maintenance of a library of master data, which is the common reference data that is used across the project (e.g. locations, terminology, glossary).

vi. Data validation – auditing and checking of information content and structure to ensure that the information contained in the CDE conforms to the agreed standards and structure.

vii. Support – offering support to the client and project team, both in terms of the information standards, and also in the use of the CDE.

The role undertaken by the PIM team will in many cases be absorbed into roles undertaken by existing project team members, for example the design manager

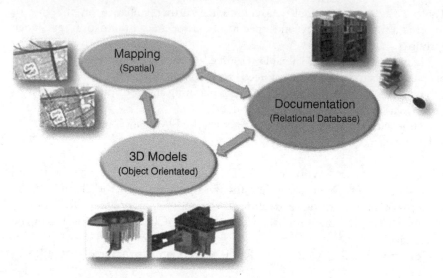

Figure 6.4 Document and data compliance (Crossrail).

on a project may well take on the responsibility for creating and managing the CDE at the early stages of a project, whereas on a larger project the tasks could be undertaken by a dedicated team, as on the Crossrail programme. The term used for the purposes of this chapter is, therefore, ubiquitous.

Case study details

The PIM team had to not only take account of Crossrail's requirements, but also had to work with both RfL and LUL as maintainers to manage their separate requirements. Both maintainer organisations use different asset management systems, which are incompatible, albeit both are working towards adopting a common system. Meanwhile, Crossrail were producing information in accordance with the latest BS & PAS1192 standards, which provided information that is significantly richer and more detailed than that previously available within TfL.

Crossrail's technical Information Management strategy was developed between 2007 and 2009. This was before the UK government launched its Construction Strategy (2011), which introduced the concept of BIM, particularly by mandating the use of BIM for all public sector projects.

In 2007/8 the programme adopted the use of BS1192:2007 as the way to ensure consultants using CAD for design could work together in a collaborative CDE, in order to ensure the spatial geometry of the infrastructure could be properly coordinated and managed. In 2010, Crossrail recognised that the document and data control processes needed overhauling to meet the future needs of the project. The principles from the British Standard were then used to create a new general document and data management and storage system using database technology. This provided robust document control functionality using workflows within the database for the different transactions required, with dropdown menus, templates and data structure providing much more rigour and control.

During 2011–2012, these same principles were applied to other areas that would traditionally have separate software applications, such as contract administration, project information request systems, snagging, asset inventory, etc. This ensured much of the key valuable information being generated – either to build the railway, or for future operation and maintenance – was being collected to provide a single source that was easy to report from and interrogate. Initially, it was difficult for some local teams and contracts to work within the Crossrail system, having historically worked to their own standards and in their own systems. However, the benefits of having one overall repository for contract and record data quickly proved to be a core backbone to the management of the project.

As discussed in further detail later in this chapter, the use of BIM, in 2009, incorporated a document and records database connected to the CAD modelling database. This meant that every published PDF drawing (either in draft or approved), was automatically stored in the records database, allowing easy access to all drawings. Crossrail also ensured that the documentation database could be accessed from their Geographic Information System (GIS).

The PIM team conducted an independent compliance review, in which Crossrail's BIM implementation and processes where reviewed against the requirements of the UK Government Construction Strategy. The review concluded that BIM delivery on the programme was broadly Level 2 compliant. It differed primarily on where specific procedures and deliverables were required, such as formal BIM Execution Plans, specified data drops, as part of new processes, and required roles, such as a BIM Manager. This arose simply because it had been necessary for Crossrail to implement its own procedures and deliverables structure well before 2011.

The Crossrail approach centred on implementing a data management infrastructure that delivered value to the business by focusing on creating an information architecture in which the core attributes were:

- common sense to ensure the right data in the right place;
- ease of access;
- security of data; and
- an enabler of collaboration.

The use of the CDE was then made compulsory within the contract documents of all design and construction contracts, requiring all suppliers to work within the Crossrail CAD Management System and Electronic Data Management System (EDMS). The GIS containing the client utility and mapping information was also made available to all designers and contractors, as another element of the CDE.

How BIM was used

Crossrail define BIM as the process of generating and managing information throughout the whole life of the asset lifecycle by using model-based technologies linked to databases of reliable information. The CRL BIM objective has been stated as:

> To set a world-class standard in creating and managing data for constructing, operating and maintaining railways by:
> - Exploiting the use of BIM by Crossrail, contractors and suppliers;
> - Adoption of Crossrail information into future infrastructure management and operator systems.

The BIM strategy sought to take advantage of new available technologies to link databases of information, together with Master Data Management (a BIM Level 3 concept), all of which gave leverage to the value and potential of information in design, construction and project management decision-making.

The BIM environment is part of this PIM world, and at its basic level comprises a set of linked databases with modern workflows, processes and procedures. Following the trend in the BIM world, federated 3D CAD models on Crossrail were kept as light and as unintelligent as possible, with all the asset details being contained in separate linked databases, enabling the asset data to be structured in a much more user-friendly and efficient way. The 3D models were easier to manipulate, for example, linked to scheduling packages to create construction sequences. So the initial core service of the PIM team on the Crossrail programme in a BIM context was to create a CDE providing a 'single source of truth' for all information enabling the future operators and maintainers to manage their assets.

Common data environment (CDE)

The key database components of the Crossrail CDE are illustrated in Figure 6.5. The 3D CAD modelling application was connected to the main document and record database so that any drawings created were automatically stored in the correct place. The GIS was also connected to both these databases, and could be used as a portal to get data and maps within the GIS or information from the document and records database.

The BIM lifecycle approach to Information Management can be recognised in Crossrail's approach to the use of the CDE. The same CDE was used by all team members through the project lifecycle providing integration of design, engineering and construction. It is intended that the same CDE will continue through into operation and maintenance in accordance with the principles of PAS1192-3:2014 (which focuses on long-term operations, maintenance and asset management), as well as decommissioning of an asset, thereby enabling the maintenance of a rich set of linked data that includes 3D models,

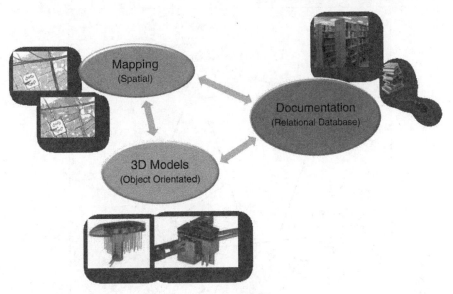

Figure 6.5 The common data environment (CDE).

spatial mapping, asset information, documentation, financials and other related information.

CRL placed as much emphasis on the graphical model itself as the information within the records database, which together give the model its intelligence.

Master Data Management

An important part of the Crossrail approach to creating the BIM (or information) environment was Master Data Management (MDM). This was necessary to mitigate the risks created by uncontrolled data. Across the UK construction industry, some estimates suggest as much as 80 per cent of vital business information is currently stored in unmanaged repositories, making its efficient and effective use a near impossibility.

The organisational needs for data and information in large infrastructure programmes such as Crossrail change and develop throughout the project lifecycle. Traditionally, this evolves with new systems and applications being brought into use alongside existing systems as and when they are required. This may ultimately result in processes and data sets that are not ideally balanced or structured because of simple incompatibility or ownership reasons (see Figure 6.6a). Crossrail needed to create the model shown in Figure 6.6b.

In 2007–8, Crossrail had a poor configuration of applications and data types, which created duplication and additional processes to enable data to be used across the project. This situation persisted until 2010 when its new information strategy was developed.

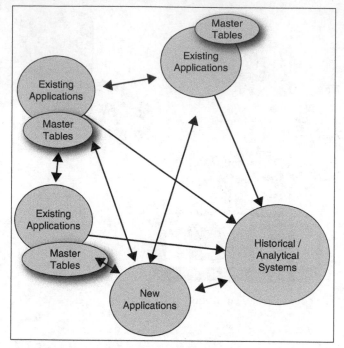

Figure 6.6a Traditional application development.

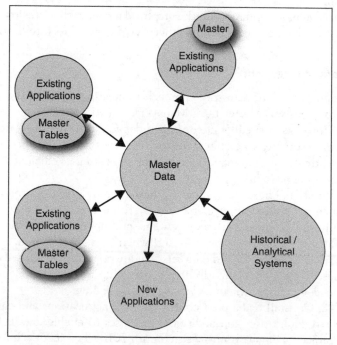

Figure 6.6b Master Data Management application development.

The PIM team, particularly those within the IT Department, worked with the Project Managers to create systems for Master Data Management. This Master Data Management enabled a data environment, in which it was possible to accurately monitor and analyse project performance as well as that of all contractors and teams. By using a data warehouse, metrics and measures are able to be consolidated across a number of systems and business areas to provide valuable reports (see also Chapter 9 Performance measurement and management).

BIM in design and construction

In 2009–10, Crossrail's Framework Design Consultants (FDCs), were contracted to undertake design work within the CDE using Crossrail's Enterprise Content Management System (ECMS) and Electronic Document Management System (EDMS) and to deliver the design in 3D and to a specified format. Their civil and structural designs were to RIBA Plan of Work Stage F (2007) (i.e. fully designed and detailed), whilst their mechanical, electrical, public health and architectural designs were to Stage E (i.e. preliminary, specification only). Crossrail created customised BS1192:2007 workflows within the CDE, which enabled the design to progress in a logical manner before being shared with other FDCs for interface coordination.

The CDE facilitated the production of integrated 3D model designs since the FDC was able to work around the constraints of the existing approved models. This approach inherently led to clash avoidance rather than having to rely on post design clash detection. Automated Quality Assurance (QA) routines were created as part of the CAD workflow, to ensure that the design data complied with the Crossrail standards before being issued. This facility was crucial in enabling Crossrail to control the vast quantity of design data being produced on the project and forms the primary function of Information Management.

The FDCs were contracted to create fully object-oriented (intelligent) 3D CAD models – i.e. model elements that know where they are (location, coordinate system), what they are (air conditioning unit) and how they relate to other objects within a system.

In 2011, construction contracts were being awarded that included significant design scope where the contractors were to become responsible for and complete the MEP and architectural designs. Crossrail developed and issued an initial 3D model 'discipline specific' level of detail (LOD) to the contractors to ensure consistent 3D model production. The contractors had also been instructed to utilise the 3D model as part of their design review process; including requirements to provide clash reports as evidence of a fully coordinated design during design assurance.

Some component libraries were also developed to support the detailed architectural and MEP designs with the intention of allowing reuse across the contracts to reduce design costs. With the industry moving towards providing standard libraries for BIM, for example BS 8541-3:2012 and the NBS National BIM library (see http://www.nationalbimlibrary.com/), Crossrail also investigated how the components produced on the project could be made available for other projects to the benefit of the industry as a whole.

In terms of project management, Crossrail used BIM modelling techniques to intelligently link the individual federated 3D model objects held within the CDE with construction tasks from their scheduling tool. The output was a timeline animation that showed the progression of construction activities over time. These were principally used in areas where there were complex interface issues or significant construction sequencing risks, and referred to here as '4D target areas'. For example, a common 4D target area identified was at the platform level of the station shafts where the tunnelling contractor needed to work in highly confined spaces alongside the station main works and enabling works contractors.

The process of creating the model was collaborative, bringing together many disciplines from the delivery teams. It enabled the project teams to explore options, manage solutions and optimise results across many target areas on the project.

The models were also used as educational and awareness tools, for daily and weekly task briefings, for safety talks and safety training.

The following are examples of how Crossrail used BIM linked to scheduling:

- Tender Support: providing visual clarity and understanding of the scope of work to be tendered, e.g. Westbourne Park and Paddington New Yard;
- Optimised Contractor Involvement (OCI): used as a collaborative tool to bring common understanding between Crossrail and contractors, in order to facilitate redesign, e.g. Farringdon Station;
- Complex Design Sequencing: used to visually represent a series of extremely complex drawings, e.g. how the Stepney Green Cavern tunnel was to be constructed;
- Interface Coordination: providing visual clarity to the interfaces between the multiple contractors involved in main stations construction and tunnelling, e.g. Liverpool Street Station, Whitechapel Station, Farringdon Station;
- Temporary Works: demonstrating the propping sequence removal within confined shafts and tunnels, e.g. Liverpool Street Station.

The Crossrail programme had a number of unique engineering challenges where 3D model visualisations provided significant insight and help. An example was to provide a timeline model of the complex sequencing for Sprayed Concrete Lining (SCL) to create the cavern at Stepney Green. This cavern was very large, at 17m high and 16.5m wide, and excavated under the East End of London. The size of the cavern meant that a complex sequencing was required and the Engineering team could not visualise how it would work from the drawings or schedules. Once a model tied these two disparate data sets together, the Engineering team were immediately able to verify the constructability of this complex work.

Infrastructure information into asset management

Crossrail developed a tagline for BIM, which was 'Crossrail are building two railways, one virtual and one physical'. The virtual railway comprises all the digital data required to manage and maintain the physical assets.

Due to the nature of the Crossrail Programme, there were a number of specific and unique additional requirements, e.g. the requirement to hand over all the technical data to the dual operators and Infrastructure Managers, which had been taken into account in order to develop a suitable approach to the provision and management of information. Careful consideration of these requirements and discussions with key stakeholders was necessary in order to ensure that the approach is both appropriate and flexible enough to allow for future needs.

Crossrail also worked with Network Rail, High Speed 2 (HS2) and LUL to try to achieve a standardised Asset Data Definition system across rail in the UK.

The Crossrail strategy provided a complete data store of information within its CDE. This comprised all the required documentation needed for acceptance of the railway by the Infrastructure Managers, along with all the relevant background digital reference and supporting data used to create the railway.

An indicative interpretation of how the CDE data store could be represented in terms of IT systems and applications is shown in Figure 6.7. The box identified

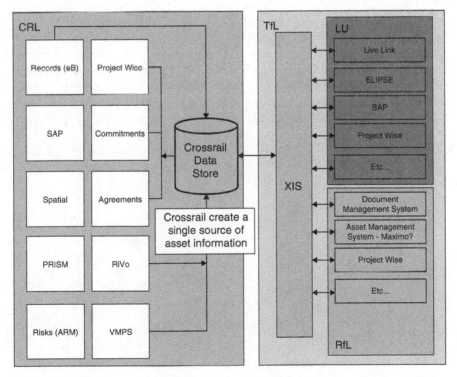

Figure 6.7 CDE becomes the data store for operations.

as 'XIS' is TfL's data integration solution. Implementation planning will need to take account of the level of maturity/usage at the time that data transfer will start to take place.

A key part of the process to hand over a new railway to the agreed Infrastructure Manager and Operator is the migration of key information packages (i.e. 3D and timeline coordination models, spatial information, O&M files, health and safety files, and certificates and asset information) required for the ongoing management, operation and maintenance of the railway.

Tools used

The Crossrail BIM environment comprises many software tools and applications, the key ones being:

- BIM authoring tool, selected as a TfL standard;
- EDMS, selected on the basis of competitive tender;
- GIS, selected on the basis of competitive tender;
- Enterprise Resource Planning (ERP) tool, selected as a TfL standard;
- Asset/Records database, selected as a TfL standard for GIS.

Issues/benefits of BIM

The way in which data is created and used collaboratively within a BIM environment is part of the fundamental shift in delivery of projects as required by the *Government Construction Strategy* (Cabinet Office, 2011). Figure 6.8 below illustrates the use of BIM thinking to coordinate all aspects of the information exchange. Traditionally, information would be kept by individual parties who continuously interact with each other, resulting in multiple copies of the same information with the concomitant problems of version control, etc., as indicated on the left-hand side of Figure 6.8. In a BIM world, data and information is sourced centrally from a CDE, controlled and distributed as required, as indicated on the right of Figure 6.8.

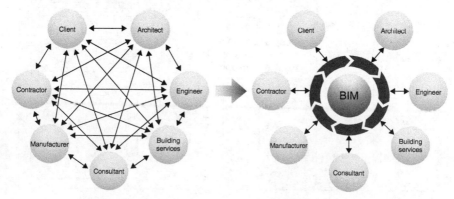

Figure 6.8 Traditional disparate data moving into a CDE.

Using BIM through the full project lifecycle has been evidenced to give better buildings and infrastructure more quickly, and at a lower cost in comparison with traditional methods of Information Management. The effective and continued use of the CDE with all project history throughout all the project lifecycle phases will lend itself to further benefits in generating long-term maintenance cost plans for the railway.

In the absence of BIM and the use of a CDE, each contract would have used its own separate data store and process for interfacing with CRL, resulting in:

- multiple CAD designs that may not spatially fit together;
- multiple data storage repositories requiring significant administration;
- major problems collecting data together at the end of commissioning and then handing it over to the future Infrastructure Managers.

Whilst difficult to calculate, these issues alone could have added many £100 millions to the cost of the project through errors and delays and so the use of BIM principles has been critical to the successful delivery of Crossrail.

The benefits of using BIM linked to scheduling to support project management include:

- Team building through sharing the visual model, which encourages an environment where challenges are reviewed and resolved in a collaborative way.
- Highlighting potential areas of safety risks and providing assurance to the construction team that performance expectations are being supported and met by all.
- Analysis and verification of construction method options prior to commencing works on site.
- Identification of schedule improvements through the re-sequencing of activities, redesign of building components and/or the reallocation of work areas.
- Identification of missing or incomplete design data.
- Mitigation of risks associated with complex construction sequencing or interfaces by providing assurance that the proposed works are viable and realistic.
- Powerful site communication and coordination allowing all project team members to quickly understand the construction site, project scope and intended construction sequences.
- A proactive team-building tool since the shared, visual model encourages an atmosphere where challenges are reviewed and resolved in a collaborative environment.

Lessons learned

The following provides a summary of the key lessons learned from the Crossrail project, and is intended to provide guidance around a core set of Information Management criteria.

The CDE

- The CDE needs to be structured with the end in mind and at the beginning of the project.
- The level of detail only needs to be commensurate with the scale and project stage.
- Revision controls, which display the current data in a manner that makes it easy to access and view the history of earlier revisions, provide great assurance to the authenticity of information.
- Document and data approvals management enables tracking progress of approval.
- Process and workflow functions include tasks, RFI's, contract correspondence and change control, actions and tracking progress against those actions.
- With a well-structured data repository, the 3D models are able to be light as reasonably practicable, albeit retaining their object orientation and being directly linked to the records database.
- Crossrail learnt to avoid packing models with data about their attributes, history etc. – this is much better stored separately in a database where you can sift, sort and form all sorts of useful relationships for operations and maintenance.

Asset classification

- Classification, coding and dictionaries are an important part of linking data sets together. The classification and coding are the lifeblood of the future maintenance system.
- Utilise Asset Dictionary Data Definition Documents to set out exactly what is required for each maintainable asset. The data required for each asset needs to be collected during construction and then stored in a records database, which can be linked to all the relevant associated operation and maintenance documentation, as well as the 3D models.

Collaboration methodology

- Greater prescriptiveness on the scale of a project the size of Crossrail (in context to the number of architects, designers, and contractors) may have yielded more predictable results and predefined the Information Management culture.
- The Infrastructure Maintainers of Crossrail require data sets that are completely integrated and unified across every station and consistent across every contract. If Crossrail had not been specific about what it wanted, the data would have been delivered in an inconsistent fashion with varying degrees of quality, and the deployment of IFC methodologies would not alleviate these issues on a project as complicated as Crossrail.

Process

- The cultural barriers to implementing BIM should not be underestimated, as there is a requirement to make significant changes to well-embedded work processes. This starts in procurement to set the right collaborative approach.
- Clear requirements, levels of detail and instructions are essential to ensure that the project requirements are understood.
- Design change should be managed within the CDE and not as an external or additional process.
- Clearer guidelines are required for use of BS1192:2007 and PAS 1192-2:2013 when issuing design for redlining, as built and schedule integrated modelling.
- Understanding the infrastructure maintainers' information requirements and involving them early in the project is critical.

People

- Adoption of BIM requires strong leadership with a clear understanding of the ultimate deliverable or 'vision of the end'.
- Collaborative working is essential to ensure all parties (client, designers, contractors and third parties) and departments (technical, commercial, engineering, procurement and construction) are pushing in the same direction for BIM adoption across the project lifecycle.
- BIM implementation will only be as strong as the weakest link in the supply chain. Training through the whole supply chain is critical, particularly in the current environment where BIM competencies are still maturing.
- The whole project team need to understand their role in BIM as it affects all phases of the project lifecycle. This needs to be supported by specialist staff, such as the PIM team and scheduling specialists who act as the glue to bind the teams together.

Technology

- Clients need to push for open standards and a move away from using proprietary formats that restrict the supply chain and stifle innovation.
- Consistent application of standards is fundamental to the success of BIM. This includes modelling standards but also standards that relate to data that links to the model, for example, a schedule Work Breakdown Structure (WBS) needed for successful timeline modelling.
- Intelligent (object-oriented) 3D models are an essential foundation for leveraging time, cost and design analysis.
- As the importance of the 3D model increases, so does the importance of model data security. Data model security within the CDE needs to be well structured, flexible and scalable.

- Ensure that the technology infrastructure (networks, broadband, etc.) is well established for site/satellite offices so that they become effective and not disengaged remote hubs.

Summary

PIM is crucial to the successful implementation of BIM on any project, with clear responsibilities and standards adopted by all members of the project team. On complex projects or programmes, this is likely to be a dedicated role, whereas on non-complex projects PIM will more likely be undertaken by the members of the project team as part of their existing roles.

The design and construction complexity of the Crossrail programme of works required that Crossrail as a client provided a collaborative and integrated environment to enable successful project delivery. The BIM methods, processes and technologies used have contributed to the design and construction works being delivered substantially on time. The BIM environment itself has been successful because PIM has been deployed on the programme.

The benefits of working within a BIM environment have already brought significant gains in terms of cost and time. Further benefits will accrue from 2016 onwards when Crossrail begins the process of asset handover to the infrastructure owners and operators.

Ultimately, BIM within Crossrail has formalised and elevated Information Management practice from being an IT-centric function to being focused on delivering genuine business value by inspiring new ways to utilise data, improve safety, mitigate risk and enrich the decision making process within programme planning by the use of logical information structures, thus enabling the collaboration of information, and, most importantly, people, reinforcing the need and value of the Project Information Manager.

References

BSI (2007) *BS1192 Part 1: Collaborative production of architectural, engineering and construction information code of practice*, British Standards Institute, London.

BSI (2012) *BS8541 Part 3: Library objects for architecture, engineering and construction: shape and measurement code of practice*, British Standards Institute, London.

BSI (2013) *PAS1192-2 Specification for information management for the capital/delivery phase of construction projects using building information modelling*, British Standards Institute, London.

BSI (2014) *BS1192 Part 4: Collaborative production of information: fulfilling employer's information exchange requirements using COBie – code of practice*, British Standards Institute, London.

BSI (2014) *PAS1192-3 Specification for information management for the operational phase of assets using building information modelling*, British Standards Institute, London.

BSI (2015) *PAS1192-5 Specification for security-minded building information management, digital built environments and smart asset management*, British Standards Institute, London.

Cabinet Office (2011) 'Government Construction Strategy'. Online. Available HTTP: <http://www.cabinetoffice.gov.uk/resource-library/government-construction-strategy>, accessed May 2015.

NBS (2015). National BIM Library. Online. Available HTTP: <http://www.nationalbim-library.com/>, accessed May 2015.

© SUE PITTARD

7   Contractual frameworks for BIM

Khalid Ramzan

Introduction

A number of industry reports (Banwell, 1964; Egan, 1998; Latham, 1998; Strategic Forum, 2007; Wolstenholme, 2009) have identified barriers to a collaborative working culture, most notably amongst the barriers are the use of bespoke contracts and outdated procurement strategies (Nielsen, 2007). In an attempt to overcome these specific barriers it is important to use contracts that allocate risk fairly and align behaviour to provide mutual benefits to all parties to the contract, for example encouraging fair and equitable payment arrangements.

When an organisation is allocated a risk that it cannot adequately control, it will seek protection (Thomsen et al, 2009) by increasing costs upfront or by engaging adversarial behaviour during the project, instigating the 'claims-based culture' that has traditionally blighted construction. Transactional contracts foresee a single result; the value of a future outcome in exchange for money (Williamson, 1979). But as Matthews and Howell (2005) state,

> The dispute record of the construction industry proves that drafting transactional contracts for the delivery of complex and uncertain construction that foresee all contingencies, allocate all risks, limit opportunistic behaviour and still motivate highest global efficiency is impossible.

A key role for the quantity surveyor (QS) is to help develop the contractual frameworks for delivering projects and then to implement and administer the contracts. This chapter addresses some of the key considerations for QSs encountering for the first time the need to incorporate BIM-related obligations into construction contracts and design appointments. The case study examines the issues to consider when incorporating BIM obligations and in particular considers an approach now available to the industry, through the use of the Construction Industry Council's BIM Protocol ('CIC Protocol'),[1] which can be incorporated into existing industry standard forms of construction contract and consultant appointments. The CIC Protocol is discussed in more detail in the Annex to this chapter. It should be noted that some organisations have developed (or are in the process of developing) their own protocols and working practices for BIM. Whilst these are outside the scope of this chapter, QSs should be alive to their existence, and although the specifics of each approach may vary, the underlying principles discussed in this chapter are likely to be common.

Author biography

Khalid Ramzan is a Senior Associate specialising in law relating to construction and engineering at the international law firm Pinsent Masons LLP. He has an MSc in construction law from King's College London where his dissertation examined the legal and strategic risks facing main contracting businesses implementing BIM. Khalid is also experienced in advising industry participants on incorporating BIM-related obligations into contracts and was a founder member of the CIC's BIM 2050 Group. Khalid regularly writes on BIM-related issues and speaks at conferences on the subject.

Company/project information

Pinsent Masons LLP is a full-service international law firm that advises project participants (procuring authorities, funders, main contractors and major sub-contractors) on domestic and international construction and infrastructure projects. Specialist lawyers in the firm regularly advise on procurement structures and delivery frameworks across sectors as diverse as commercial property development, solid waste processing, nuclear (newbuild and decommissioning) and healthcare. In recent years, the firm has begun to advise project procurers and main contracting businesses on the legal issues arising from the use of BIM.

Service Profile

Core service provided by the QS

In the construction industry there are a series of pre-existing contractual frameworks, which can be adapted in practice to record the relationships that are formed for the delivery of construction work and related services. These include the most common frameworks issued by standard form publishers widely recognised in the industry, such as JCT, NEC and, more commonly on international infrastructure projects, FIDIC. Each of these organisations publish a range of contractual conditions suitable for particular types of risk allocation and pricing structures. For example, a form of contract such as the JCT Standard Building Contract[2] imposes an obligation to construct on the basis of design carried out by consultants engaged by the client. In contrast the JCT Design and Build Contract[3] form places design and construction risk on the contractor.

Forms of contract often also differentiate on the basis of *how* pricing has been agreed under a contract. The NEC ECC (Engineering and Construction Contract) is a good example of this. Its suite of Main Options ranges from

Option A, which is a form of fixed price contract, through to Option E, which has a reimbursable contract pricing structure.

In summary, before factoring in BIM, some of the fundamental questions that the QS must ask to identify the most suitable standard form for a particular project include:

- Will the contractual relationship be at main contract or subcontract level?
- Will the supplier be taking design or construction risk, or both?
- What will be the commercial basis for remuneration under the contract? For example will it be fixed price lump sum, reimbursable, or will there be a target price with a pain/gain share?

The need to identify the most appropriate standard form can raise a complex set of issues for the QS to consider. Once the parties to the construction project have agreed on the contractual basis for their relationship, the QS will be charged with putting together the contract for parties to sign. Commonly, this will require agreement between the parties on such matters as the scope (in the Preliminaries or the Employer's Requirements, or the schedule of services or scope in the case of professional services), the contract documents, method statements and the like.[4]

In addition, specific projects will give rise to their own complications. For example, in the case of commercial property development projects, there may be a number of stakeholders who will require rights of recourse to those responsible for design (through collateral warranties, for instance).

A key role of the QS prior to contract signature is to ensure that the documents appended to and properly incorporated into the contract reflect the true scope of what is required and the nature of the intended contractual relationship. Contract documents may include the Employer's Requirements or other essential details of the outputs that the client expects.

Other documents incorporated might require the design to be developed to a particular stage against which the contractor must build (for example by reference to the work stages of the RIBA Outline Plan of Work 2013). In addition there are likely to be method statements which would represent the contractor's response to how the works will be executed.

Additional considerations for the QS on BIM-enabled projects

The advent of BIM raises a fresh set of issues to consider. The QS must consider the additional provisions to be made in the contractual documents to address responsibilities, risks and liabilities relating to digital modelling and processes such as information exchange. Contract documents must define the nature and extent of the modelling obligation that the client expects to be undertaken and the processes (such as those relating to information exchange) which must be implemented to ensure that the use of BIM achieves the client's intended outcomes. In addition, concerns about the ownership of data and wider intellectual property (IP) need to be addressed.

One approach to framing BIM-related obligations is the use of a combination of a 'BIM protocol' document, which is incorporated into the underlying contract (such as a main building contract, a consultant appointment or a sub-contract),[5] and amendments to the contractual terms and conditions. The use of the CIC Protocol in this context is discussed in the Annex to this chapter. Whatever approach is adopted, the QS must consider the contract as a whole so that when seeking to incorporate BIM obligations the scope, or similar documents, need to be reviewed alongside the terms and conditions.

Key considerations and the role of the Employer's Information Requirements

The signed contract normally represents the crystallisation of a relationship developed from the point the client issued the invitation to tender (ITT) documents, or when the particular contractor or consultant first discussed the project with its client.

Ideally, the client should already have been developing its requirements for BIM in advance of issuing the tender documents. In particular, a number of areas highlighted below should be considered part of the Employer's Information Requirements (EIRs). EIRs are used to define clearly the client's desired BIM outcomes. They should be developed prior to tendering the project team roles. Selection criteria applied during the tender process can then be used to select a project team with the best aptitudes and experience to deliver on the client's desired outcomes.

The implementation of an industry-wide set of standards for EIRs is a key challenge for the industry. Despite the efforts of several institutions, there is still little guidance and training for client-side practitioners on developing robust, but realistic and achievable EIRs. It may, therefore, mean in practice that contractors and consultants are actively engaging with clients to develop such documents. A client may elect to engage a specialist BIM consultant as it develops its procurement strategy in order that the EIRs develop in parallel. The EIRs remain central to the successful outcome of a BIM modelling exercise.

The key areas for the QS to consider when seeking to incorporate BIM-related obligations into contracts are outlined below.

The need to clearly define the outputs and information exchanges required by the client

If the client has developed the EIRs and has clearly communicated them in the ITT, this could be a fairly straightforward exercise when it comes to entering into the contract. The client's requirements may be a very general statement of outputs, or something more prescriptive. For example, the requirement could include prescribing the type, extent and quality of data to be delivered to the client on completion of the services or the 'data drops'[6] that are expected at a particular stage of the modelling process.

As with any aspect of design or construction, a key element of any contractual relationship is ensuring clarity between the parties on exactly what is to be delivered under the contract at what stage. The purpose for which a model is required will determine its levels of detail and complexity at a particular stage in the design/construction process. For example, a model required solely for the purposes of identifying design clashes between the steel structure and mechanical and electrical engineering systems will not need to contain the level of detail that would be required in a model that is capable of being exploited to predict thermal efficiency during the asset life. See in the Annex to this chapter how the concept of levels of detail is adopted in practice for the CIC Protocol.

Therefore, central to any BIM contractual framework is defining the type of modelling required (by reference, for example, to industry standard levels of detail) at any particular stage of the design or construction process. Generally, a project is more likely to successfully deliver the client's desired BIM outcomes if the client has invested in developing its requirements in detail rather than in general terms.

The need to define the agreed processes to be adopted for implementing BIM

This should ensure that a single set of processes can be applied in a uniform manner (including through other contractual relationships which the client will have with other consultants on the same tier, and relationships between consultants or contractors and their respective subcontractors) to bind all parties involved in modelling and information exchange. This will help ensure, for example, that the parties working in a BIM environment are compelled to work within an agreed common set of file formats, processes and standards so that the required outputs, information sharing and reuse can be achieved. Again this should form a fundamental part of the EIRs upon which the client should go out to tender, and these detailed requirements (which may have developed further between tender and appointment) can be incorporated into the scope.

The base contractual obligation to carry out modelling should be properly incorporated into the contract

This can be achieved by means of a suitably drafted clause in the terms and conditions part of the contract referring to the modelling obligations. Such obligation may already be captured in the contract terms by a simple statement that the contractor or consultant must carry out the works or services in accordance with the scope as defined. For example in the NEC 3 PSC (Professional Services Contract)[7] the base obligation at clause 22.1 provides that 'the *Consultant* Provides the Services in accordance with the Scope'. If the Scope is the document in which the modelling obligations are to be incorporated (as the technical requirements should be) no further amendment will be required. However, it is worth checking that in any given set of contract conditions the fundamental obligation to carry out modelling has been properly incorporated.

Standard form contract publishers have been slow in responding with their own guidance on how the CIC Protocol, or indeed any BIM protocol, should be incorporated into their forms. Some early guidance from the JCT in 2011[8] suggested incorporating 'any agreed Building Information Modelling protocol' into the 'Contract Documents' (as defined under the particular JCT form) without, seemingly, much thought given to what the protocol may actually comprise.

In 2013 the NEC as part of the re-launch of its suite of updated NEC 3 forms of contract, published a 'how to' guide for BIM.[9] This guide is, arguably, the most significant guidance by a contract publisher to date on the practicalities of incorporating any protocol, but with particular reference to the CIC Protocol. Referring specifically to the CIC Protocol, it suggests that the approach to incorporation should be to 'split' the CIC Protocol into the terms which are incorporated as 'Z clauses' (i.e. amendments to the relevant standard form terms) with Appendices 1 and 2 of the CIC Protocol incorporated into the Works Information or Scope. The rationale for this approach, NEC argues, is:

> The *Project Manager* (ECC) or *Employer* (PSC) is able to alter the technical requirements by instructing a change to the Works Information or Scope, such as including any revisions to a published protocol. He is also able to instruct the *Contractor* or *Consultant* on how to deal with any ambiguity or inconsistency which arose between the protocol requirements and other sections of the contract.[10]

Ownership and rights to future use of data within a BIM model

The client should consider, when developing its EIRs, the extent of control and ownership it requires over the intellectual property (IP) that is created through the implementation of BIM on its project. The extent of such ownership and control will be determined in part by the outcomes the client intends to achieve from BIM and, in part, from what is commercially achievable in the market. For example, the client may intend to use the data captured through BIM to support operational use and facilities management. It may, therefore, conclude that it needs to own the data created by its consultants and contractors. The supply chain is likely to resist demands to pass ownership of data to the client and may insist on the more familiar approach of licensing the data. Designers are likely to have their own concerns about the conditions on which data they create is licensed and used on the project and the responsibility they might bear to the client or third parties for its long-term use.

IP is typically addressed in the terms and conditions of contract. For example by setting out the terms of the licence given to the client for future utilisation of the software used to undertake the modelling and/or the licence under which the client can use the output of the modelling (the data and any reports created from that data). This issue is addressed in the Annex to this chapter in the context of the CIC Protocol. The CIC Protocol takes a very particular approach to IP, which is licensed by creators for a 'Permitted Purpose' (as defined) and this may not be suitable for all projects.

Whilst common standard form contracts address IP, they largely pre-date the widespread adoption of BIM. The JCT forms for example provide that: 'Subject to any rights in designs, drawings and other documents supplied to the Contractor for the purposes of the Contract... the copyright[11] in all Contractor's Design Documents shall remain vested in the Contractor'[12]. On any given project, the use of BIM may give rise to issues of IP ownership such as database rights, joint ownership of IP and terms of licensing. Whilst such issues should not be unduly complicated in an environment where discipline-specific models are created for particular purposes, clients and suppliers should seek specialist legal advice particularly to ensure that ownership of particular types of IP, such as databases, reflects the intention of the parties. Legal advice may also be appropriate where specific concerns arise around the ownership of IP created through modelling, particularly when participating in a BIM modelling exercise for the first time.

Allocation of liabilities arising from modelling and responsibility for errors in the models produced

This should also be addressed in the terms and conditions. This is an issue of substantive risk allocation (and therefore best addressed in the terms and conditions), but the industry consensus seems to be that BIM at its current stage of development will not entail any great alteration in the ways in which design liability, and liability for information provided under contracts, will be allocated.

Certain other issues, typically of a contractual nature such as clarity around what aspects of the overall contract take precedence over others (often referred to as 'priority of documents')

These can also be addressed in the terms and conditions. The key concern here is to ensure that in the event of a conflict between the terms of the contract and the contents of the scope documents, there is clarity over which parts of the contract take precedence. See the discussion in the Annex to this chapter on how such issues are addressed in the CIC Protocol. In a BIM context the parties may intend that certain aspects of the BIM protocol take priority over terms and conditions. To achieve this, the priority clause should be drafted to reflect such intentions.

The role of BIM information management can be identified

This role should be defined, together with its obligations and responsibilities (see Chapter 6 for more on the role and function of BIM information management). The details of what that role entails would be defined in the scope document for the appointment of that role. However, the obligation on other project parties to co-operate and comply with information management, could be detailed in the contract terms and conditions by means of a simple contractual obligation (if not already addressed in the scope).

Incorporating the BIM obligations into existing contracting structures

The number of issues discussed above for consideration in a BIM context may suggest that current contracting practices will need to be significantly altered to accommodate the wide range of additional issues that the contract must address for BIM. However, that is not necessarily the case. The approach taken in the industry thus far has been to assume that existing two-party contractual relationships will not be radically overhauled. Rather, institutions and organisations such as major contractors, have largely adapted current industry standard forms in order to accommodate BIM-related obligations.

In addition, some standard form publishers have developed guidance and a limited set of new documents to better enable working in a BIM environment. In some cases, there has also been a certain limited amount of standard contract drafting to address the legal issues that are perceived to arise.

This approach is in line with a level of consensus that has emerged within the industry which suggests that with BIM at its current stage of development, existing processes and legal frameworks do not have to radically alter. This view emanates from the conclusions in a report in 2011 by the Government Construction Client Group[13] which found that: 'little change is required to the fundamental building blocks of copyright law, contracts or insurance to facilitate working at Level 2 of BIM maturity'.[14]

Impact on traditional processes

Contracting structures such as those developed by JCT and NEC, which predate BIM, are likely to remain in place for the foreseeable future. This means that one very fundamental feature – the fact that such contracts are two-party relationships (bilateral) between a client and a supplier – will remain fundamental to the contractual relationships that accommodate BIM. However, achieving the full potential of BIM requires collaboration, and that potentially creates dependencies between suppliers on the same tier that are likely to be more critical than in non-BIM environments.

For example, in developing a base reference model, the architect and structural engineer may have to ensure that they carry out their respective tasks in a particular pre-agreed sequence and produce information in a standardised format adopting standard naming conventions so that the output can be compatible with models produced by other members of the design team. The timely coordination of the mechanical and electrical engineer's model with the structural model, would be critical in order to avoid clashes.

If one of the design team members does not adhere to the standardised processes and naming conventions, the resulting errors (in this case failing to identify design clashes virtually so that they are only picked up on site) could cause project-wide difficulties (i.e. the realisation of clashes during construction and the ensuing time and cost implications of that). This would undermine the aims of adopting greater collaboration. In addition, where a particular discipline does not adhere to the processes necessary for collaboration, this could potentially cause other individual disciplines significant volumes of rework, leading to additional fees and ultimately a more expensive project.

However, for now, suppliers on the same tier are unlikely to have any contractual recourse directly to each other, and will continue to have their direct relationships with the client or the main contractor who will co-ordinate the relationship between two or more suppliers on the same tier.

For good or ill, the industry seems to have approached the greater need for collaboration by strengthening processes and working practices. Standards such as PAS 1192-2:2013[15] have been developed specifically to improve collaboration through the use of standardised conventions, procedures and working practices. There has been less focus on developing genuinely multi-party contracting structures under which parties of the same tier owe obligations to each other.

Tool capability

There is currently a lack of any specific tools to support the development of the contractual framework for BIM Level 2-complaint projects, other than in the form of supporting underlying process through workflow functionality typically provided as part of document and information management systems.

Issues/benefits

When finalising contracts, QSs should consider any necessary modifications to the terms and conditions of the underlying contract, and the contract conditions should, therefore, be reviewed to ensure that the following issues in particular are adequately addressed:

- The base contractual obligation to carry out modelling.
- The ownership and future use of IP created from modelling.
- A priority clause clarifying which aspects of the contract take priority in the event of inconsistency.
- Provision to support the role of information management.

In addition, consideration should also be given to the following:

- The number of amendments and/or addenda to be incorporated into the underlying contract. This is particularly important where parties elect to make bespoke changes in preference to adopting a standard protocol.
- Issues related to risk profile such as the duty of skill and care, and the extent of associated modelling obligations.

- Issues related to levels of detail and intended use thereof, i.e. to limit the use of model detail to the purpose(s) intended.

Always consider the use of standard protocols before electing to draft bespoke amendments.

Summary

Working in a BIM-enabled environment and delivering digital modelling should be a contractual obligation, as clear and precise as any other contractual obligations taken on by members of a construction supply chain. The key to framing BIM obligations is a properly defined and priced scope. This is why the EIRs take on a central importance as they should be clear and achievable.

The use of BIM, however, differs from other services and works delivered on a project, because the successful delivery of digital modelling requires parties involved to adhere to common processes and standards in the delivery of their respective outputs. This is why, when preparing contracts, the QS must have regard to the entire supply chain on the same tier. Whilst the contractors and the designers may be appointed by the client and have no formal contractual relationship with each other, the processes incorporated must be sufficiently uniform to enable members of the project team to interface adequately in a collaborative BIM environment.

One effective way of defining both the scope of BIM-related obligations and the processes that the supply chain should adopt, has been the use of BIM protocols, which can be appended to existing contract forms as part of the scope without the underlying contract needing to be radically rewritten. That said, it is always advisable to seek the appropriate legal advice before considering any modifications to the terms and conditions of the underlying contract.

Annex

The CIC BIM Protocol ('the CIC Protocol')

This Annex applies the issues outlined in this chapter to the use of the CIC Protocol, as the first UK industry-wide form of BIM protocol and guidance.

The CIC Protocol has been developed to fit into the bilateral contracting structure described above. In order to understand how the CIC Protocol could be used to amend a scope of services/specification, and thereby incorporate BIM-related obligations, it is necessary to understand some of the key features of the CIC Protocol.

Outline of the structure of the CIC BIM Protocol

Parties with modelling obligations are defined in the CIC Protocol as Project team members ('PTMs'). The CIC Protocol is intended to become part of a contract, such as a design appointment or construction contract (that underlying contract is referred to in the CIC Protocol as 'the Agreement'). The CIC Protocol is intended to be incorporated into all direct contracts between the client and members of the project team. So, for example, on a project where the client employs separate design consultants and a building contractor, the CIC Protocol is to be incorporated into individual design team appointments as well as the building contract.

The CIC Protocol consists of three parts:

- A set of contractual conditions which are expected to form part of the Agreement.
- Appendix 1 – the Specimen Model Production and Delivery Table.
- Appendix 2 – the Information Requirements.

The set of contractual conditions which are expected to form part of the Agreement consists of eight clauses, these are:

- Clause 1 – a set of definitions that apply to the CIC Protocol and (potentially also) to the Agreement.
- Clause 2 – a priority clause which provides that in the event of inconsistency between the terms of the Agreement and the CIC Protocol, the CIC Protocol prevails unless the CIC Protocol states otherwise.
- Clause 3 – the obligations of the employer under the CIC Protocol.
- Clause 4 – the base obligations of the PTM in relation to modelling.
- Clause 5 – a statement that PTMs do not warrant the integrity of any electronic data delivered in accordance with the CIC Protocol.
- Clause 6 – a clause addressing the use to which models produced under the CIC Protocol may be put, including ownership of any IP in them.
- Clause 7 – the liability of a PTM for the models it produces.
- Clause 8 – what terms of the CIC Protocol are intended to continue to have effect following termination of the PTM's employment under the Agreement.

The base modelling obligation

The base obligation in the CIC Protocol is for the PTM to produce models ('Specified Models') to the 'Level of Detail' specified in the 'Production and Delivery Table' for a particular 'Stage' of the project (clause 4.1.1). All of these terms are defined in Clause 1 of the CIC Protocol and it is important for QSs to be familiar with their precise meaning in the context of a particular contract. The meaning of all these terms may differ significantly depending on the way they are used in the relevant Appendix.

The base obligation in Clause 4.1.1 is arguably the most important aspect of the CIC Protocol, aimed at defining the full extent of the PTM's modelling obligation.

Levels of detail

A key concept in the CIC Protocol is defining the level of detail or complexity to which a model should be developed by a particular stage. This is an element of the CIC Protocol that requires engagement by the client's technical BIM advisers as this is the opportunity to ensure that the extent and complexity of modelling that the client expects matches the objectives the client has identified for using BIM on its project.

In Appendix 1, the CIC Protocol uses the concept of levels of detail (LOD), which defines the stage to which a model is to be developed and the CIC Protocol links this to the levels of reliance that can be placed on a model developed to that particular LOD (see the section on levels of reliance later in this Annex for more information). Significantly, LOD is also linked to the concept of 'data drop'.[16]

Appendix 1 contains a pro-forma 'Specimen Model Production and Delivery Table' in which each Model can be defined using references to work stages (e.g. brief, concept, developed design, etc., the referencing conventions of PAS 1192-2:2013 being used for this purpose, although the guidance to the CIC Protocol makes clear this is only a suggested approach). For each LOD each Model is to be assigned a 'Model Originator' who will be responsible for its development at the particular work stage. The Model Originator for a particular model may be another supplier – the CIC Protocol is intended to be replicated in all design appointments/contracts on a project.

The CIC Protocol is not prescriptive in terms of the model definition conventions to be used. Although the referencing conventions of PAS 1192-2:2013 are used to define the required level of design development, they are used alongside the Association of Project Managers (APM) design stage definitions. The QS will have to consider the most appropriate conventions to adopt on a project so that they can be readily understood and utilised by the whole project team. It may be, for example, that the RIBA Plan of Work 2013 will be adapted for this purpose as a universal convention system in the future.

Appendix 2 as the vehicle for defining the practical arrangements for modelling

The QS must take care to define, in Appendix 2, all the necessary 'ground rules' to ensure that modelling for a given project can be successfully undertaken. Clause 4.1.2(c) of the CIC Protocol requires the PTMs to comply with the 'Information Requirements'. 'Information Requirements' are defined as 'the document attached to this CIC Protocol at Appendix 2 setting out the way in which Models shall be produced, delivered and used on the Project, including any processes, protocols and procedures referred to therein'.

The CIC Protocol guidance indicates that the framework presented in Appendix 2 is generic and the QS may choose to introduce further types of information if they require. However, as the definition of Information Requirements implies, Appendix 2 is the place in which all the documents defining the *process* of modelling should be located. It is through Appendix 2 that one of the key requirements of successfully implementing BIM, namely standardised processes and procedures, can be defined. The key issues that Appendix 2 must cover are:

- The standards that are to apply to the modelling process (e.g. PAS1192-2:2013 or similar). This includes the software that is to be used to deliver the project, the file format for the delivery of models and any applicable naming conventions.
- Identification of all the parties involved in the modelling exercise on the project.
- Identification of the party that will take on the role of information management at any given stage of the project.
- A description of the EIRs (in effect the key elements of the information outputs that the client expects on a project – the role of the EIRs is outlined earlier in this chapter, together with a brief overview of EIRs and considerations when developing them).
- Identification of other applicable project procedures, namely any additional protocols that might apply, for example, in relation to data security.

The CIC's guidance notes accompanying the CIC Protocol envisage Appendix 2 to be based essentially on the BIM-related outputs identified by the client in the project tender documents.[17] It is therefore essential that the client's BIM advisers are active at the tender stage in assisting the client to develop and define the client's required BIM outputs. This in turn will influence the processes the contractor or consultants will adopt to deliver those outcomes. See the section on the role of the EIR outlining how important this is in developing the client's desired BIM outcomes from an early stage, and how it can aid the client in selecting a project team that has the best potential to achieve the desired outcomes.

The CIC also envisages Appendix 2 to be subject to development during design and construction, and is, therefore, deliberately intended to be an evolving document.[18]

The role of Information Manager is highlighted in the CIC's guidance and the CIC has published a short specimen scope of services for that role, which can be used as the basis of the appointment document for an Information Manager.[19] Whilst the role of information management is outside the scope of this chapter (see Chapter 6 for more on the role of information management), one of the duties envisaged by the CIC Protocol is for the Information Manager to review the contents of Appendix 2 at regular intervals during the course of the contract to ensure that it remains relevant to the on-going modelling requirements. Changes to Appendix 2 are envisaged to be subject to the change control procedure in the underlying contract, but this depends to some degree on how the CIC Protocol is incorporated into the underlying Agreement (in the section below headed 'Effectively incorporating the CIC Protocol into the contract documents' there is a discussion on incorporating the CIC Protocol into NEC contracts).[20]

In what circumstances should the CIC Protocol be adopted?

The CIC Protocol is essentially a 'path-finder' document that has been developed in recognition of the fact that BIM was still relatively new to the UK industry. The guidance in the CIC Protocol acknowledges this, with one of its stated objectives being to

> support the adoption of effective collaborative working practices in Project Teams. The encouragement of the adoption of common standards or working methods under PAS 1192-2:2013 are examples of best practice that can be made an explicit contractual requirement under the Protocol.[21]

There are examples of other protocols being developed in the UK industry, in particular by some of the major main contracting businesses, however these have been developed for internal use by supply chains on major projects. It is likely the CIC Protocol will remain as an industry starting point whenever BIM-related obligations are to be incorporated contractually.

As stated earlier, the CIC Protocol is intended to be incorporated into all direct contracts between the client and project team members in identical form. On design and build projects, the QS should carefully assess how the obligations will fit together, particularly following novation of a design consultant to the building contractor who would originally have been appointed by the client. The CIC Protocol guidance anticipates this being an issue and advises:

> On a Design and Build project, it [the CIC Protocol] will initially be appended to contracts of the design team entered into prior to appointment of the Contractor. When the Contractor is appointed, the Building Contract should make him responsible for providing the models and should include the Protocol. If all consultants are novated, the Protocol appended to the novated appointments will allocate responsibility between members of the design team. If some consultants are not novated, careful consideration should be given as to how the responsibilities under the Protocol will be allocated following novation.[22]

On a project where novation is likely, the QS must be particularly careful to ensure that the CIC Protocol is amended appropriately so that it works effectively post novation. Issues of particular concern are likely to be ensuring that IP licences post novation are in favour of the contractor as well as the client (if following novation the client also requires rights in IP as well). Appropriate legal advice should be sought to ensure that novation has the effect that the parties intend.

Incorporating the CIC Protocol in the construction contract and design appointments

The client is obliged to ensure that terms identical to the CIC Protocol are incorporated into all contracts (including sub-contracts) and design appointments of parties with modelling obligations. Clause 3.1.1 provides for the employer to 'arrange for a protocol in substantially the same terms as this Protocol ... to be incorporated into all Project Agreements'. In addition, the QS should ensure that where any parts of the project are to be delivered through subcontracting, then PTMs must 'arrange for this Protocol to be incorporated into any sub-contracts that it enters into in relation to the Project to the extent required to enable the Project Team Member to comply with this Protocol' (Clause 4.1.3).

Issues and benefits

Key issues and benefits

Some of the key issues and problems that need to be addressed when the use of the CIC Protocol is contemplated are outlined below. These are:

- How the CIC Protocol is actually incorporated into the underlying contract.
- How the CIC Protocol addresses priority of documents.
- Issues with the risk profile within the CIC Protocol including:
  - The duty of skill and care.
  - The extent of the modelling obligation.
  - The IP licence dependent on fees having been paid.
- Levels of reliance to be placed on models.

Effectively incorporating the CIC Protocol into the contract documents

A key issue that will confront the QS is how the CIC Protocol must be incorporated into a construction contract or design appointment. As discussed above the CIC Protocol consists essentially of two parts. The first part is the eight clauses similar in concept to any set of terms that will govern the relationship between client and contractor. The second covers the technical requirements that are described largely in Appendices 1 and 2.

The guidance to the CIC Protocol states that the CIC Protocol is to be 'appended' to the underlying contract.[23] The CIC Protocol provides a 'Model Enabling Amendment' to achieve this.[24] It envisages that the CIC Protocol will be attached as an appendix. It also sets out a short clause which provides that the employer and contractor will 'comply with their respective obligations set out in the BIM Protocol{...}have the benefit of any rights granted to them ... and have the benefit of any limitations or exclusions of liability contained in the BIM Protocol'.[25]

The guidance also makes clear that actual incorporation of the CIC Protocol needs to be considered on a contract-specific basis and 'legal advice should be sought in this regard'.[26] Of the standard form publishers, NEC has issued guidance on how the CIC Protocol should be incorporated into its suite of contracts. NEC's suggested approach recognises that the technical requirements of modelling, such as the levels of detail, the extent of modelling required (Appendix 1) and the process of modelling (Appendix 2), if they are altered following contract commencement, should entitle the supplier to claim a compensation event. These should therefore be incorporated in the scope, whereas aspects of the CIC Protocol that comprise contractual terms and conditions should be incorporated by way of 'Z clauses'.

Terms of the CIC Protocol prevail over the contract to which it is appended

If there is any inconsistency between the terms of the CIC Protocol and the Agreement, then the terms of the CIC Protocol are to prevail (Clause 2.1) 'except where the Protocol states otherwise'. The QS should be very careful to review the terms of the CIC Protocol against the terms of the Agreement to ensure that anything in the CIC Protocol is indeed intended to prevail over the Agreement.

The CIC Protocol as a 'supplier friendly' document

The QS should be aware that the CIC Protocol is not 'one size fits all'. In its nature, being a template starting point, it cannot cater for every eventuality on every unique project. One key observation about the CIC Protocol has been that it is very much a supplier-friendly document. Where acting for a client, the QS must be careful to ensure that the client is content to contract on the form that is presented, or alternatively, whether the CIC Protocol terms should be amended so as to make it more 'client-friendly'. Some of the key 'supplier-friendly' features of the CIC Protocol are identified below, with suggestions on how the CIC Protocol could be adapted to rebalance it in favour of the client.

Using reasonable endeavours to model

The need to formulate the PTM's base obligation to model within Appendices 1 and 2 has been discussed above from a technical perspective. The standard of skill and care to which the PTM must carry out its modelling obligations is that set out in the Agreement ('using the level of skill and care required under the Agreement' (Clause 4.1.1)). However, the obligation on the PTM to actually carry out modelling and deliver the models is qualified to one whereby the supplier must use 'its reasonable endeavours' to deliver the models. Further, the base obligation is subject to events outside the PTM's reasonable control, including any acts or omissions of the employer, other PTMs and any third party (excluding the PTM's own sub-contractors, Clause 4.1).

From a client's perspective this would appear to be an odd way of instituting what is, after all, the PTM's central obligation under the CIC Protocol – namely to produce models. The supplier is unlikely to have an obligation under the Agreement to use its 'reasonable endeavours' to carry out its other obligations, for example to carry out design. Clients may wish to delete such 'reasonable endeavours' qualifications to make modelling obligations consistent with the supplier's other obligations, under the Agreement.

IP licence dependent on fees having been paid

Another indication of the supplier-friendly nature of the CIC Protocol is the licence which the PTMs provide the client under Clause 6. Clause 6.4 provides that a licence granted by the PTM 'may be suspended or revoked in the event of non-payment to the extent that any licence in the Agreement provides for such suspension or revocation'. Therefore, if the conditions in the Agreement link the payment of fees to the grant of the IP licence, then a licence granted under the CIC Protocol could similarly be suspended. The QS will need to be particularly wary of this. The revocation by the PTM of a licence to the client to use its models may have a very disruptive impact on the modelling exercise.

Level of reliance to be placed on models

A key concern in the industry, particularly on the supply side, has been how the levels of reliance that can be placed on a model at each stage will be defined. This is important because clearly a model created at an early design development stage for the purposes of project phasing and overall project duration cannot be relied upon for more detailed construction programming. The CIC Protocol addresses this issue by limiting the PTM's liability to the employer, other PTMs or any third party for a Model being modified, amended, transmitted or copied (Clause 7.2) to the extent to which the copyright licence is provided in Clause 6.

Clause 6 provides a licence to certain parties to use the 'Material' for the 'Permitted Purpose'. The definition of 'Permitted Purpose' is linked to the purpose for which a model is prepared and the LOD to which that model is prepared. The CIC Protocol thus defines 'Permitted Purpose' as 'a purpose related to the Project (or the construction, operation and maintenance of the Project) which is consistent with the Level of Detail for the relevant Model (including a model forming part of the Federated Model) and the purpose for which the Model was prepared'.

A summary of the key benefits that result from the use of the CIC Protocol

The benefits of using the CIC Protocol as a way of incorporating BIM-related obligations and addressing BIM-related risks can be summarised as follows:

- It is issued by an industry-wide body (although arguably one that has a particular remit for representing the viewpoint of design consultants), and is widely available and can therefore be adapted for use across the industry.
- The CIC Protocol provides an 'off-the-shelf' starting point for the QS coming to BIM for the first time, and accompanying the CIC Protocol there

is guidance on how to properly incorporate the CIC Protocol into the underlying contract and to complete it.

- The CIC Protocol does not require a radical over-haul of the construction procurement process, or construction contracting terms, it merely requires that a substantially identical CIC Protocol is entered into in each building contract and by all design appointments on a project.
- The CIC Protocol provides the QS with a focus, and prompts for thinking about the key issues, thereby maximising the prospect of issues being addressed up-front.

Summary

Any QS approaching BIM for the first time needs to be aware of the CIC Protocol and how it can be adapted to frame BIM-related obligations of a contractor or design consultant on any particular project. The CIC Protocol pro-forma can only ever be a starting point for the QS. The key elements of the CIC Protocol are Appendices 1 and 2, which form the core documents in which the modelling obligations are to be described. Once these have been settled, the QS must take particular care in ensuring that the additional terms of the CIC Protocol which relate to the BIM-related obligations remain consistent with the underlying building contract or design appointment, and truly reflect the intent of the parties.

Notes

1 The CIC BIM Protocol was launched in February 2013 and is part of a suite of documents comprising: (1) 'Best Practice Guide for Professional Indemnity Insurance when Using Building Information Models' first edition 2013 (CIC BIM/INS); (2) 'Building Information Model (BIM) Protocol' first edition 2013 (CIC/BIM Pro); (3) 'Outline Scope of Services for the Role of Information Management' first edition 2013 (CIC/INF MAN/S).
2 For example the Standard Building Contract with Approximate Quantities 2011 (SBC/AQ 2011).
3 For example the Design and Build Contract 2011 (DB 2011).
4 In this chapter, for convenience, documents detailing what is to be delivered and how under a contract are referred to generically as 'scope' whether referring to construction contracts or professional appointments.
5 In what follows in this chapter, references to contracts apply to contracts of any tier and design appointments, unless stated otherwise.
6 'Data drops' are discussed further at footnote [16].
7 2013 edition.
8 The JCT Public Sector Supplement (2011).
9 NEC 3, 2013 How to use BIM with NEC 3 Contracts (2013).
10 'How to use BIM with NEC 3 Contracts' published by NEC, April 2013, p 3.
11 Emphasis added.
12 See for example Clause 2.38 of the JCT Design and Build Contract 2011.
13 'A report of the Government Construction Client Group: Building Information Modelling (BIM) working party strategy paper' March 2011 (available from www.bimtaskgroup.org).

14 Ibid, see para 5.2 of page 6.
15 'Specification for information management for the capital/delivery phase of construc-
    tion projects using building information modelling' published by British Standards
    Institution available at www.shop.bisgroup.com.
16 'Data drops' are an essential element for defining the outputs that the client will
    require from BIM. Although the term is not defined in Clause 1 of the Protocol, a data
    drop is the information to be provided to the client at a particular stage. See also the
    guidance at item 3.8 of Appendix 2.
17 BIM Protocol – Introduction and Guidance para 5, p. vii.
18 BIM Protocol – Introduction and Guidance para 5, p. vii.
19 BIM Protocol – Introduction and Guidance para 4, p. iv.
20 See Chapter 6 on the wider discussion on the role of information management on
    BIM-enabled projects.
21 BIM Protocol – Introduction and Guidance 3.1, p. iv.
22 BIM Protocol – para 3.2, p. v.
23 The CIC BIM Protocol Guidance para 3.2, p. v.
24 The CIC BIM Protocol, Model Enabling Amendment at p. viii.
25 The CIC BIM Protocol, Model Enabling Amendment at p. viii.
26 The CIC BIM Protocol, Model Enabling Amendment at p. viii.

References

Banwell, H. (1964) *Report of the committee on the placing and management of contracts for
    building and civil engineering work*, HMSO, London.

BSI (2013) PAS1192-2 Specification for information management for the capital/delivery
    phase of construction projects using building information modelling, British Standards
    Institute, London

Construction Industry Council (2013) *BIM Protocol*, Construction Industry Council,
    London.

Egan, J. (1998) *Rethinking Construction: Report from the Construction Task Force*, Department
    of the Environment, Transport and Regions, London.

Latham, M. (1998) *Rethinking Construction: The Report of the Construction Task Force*,
    HMSO: Norwich, UK.

Matthews and Howell (2005) 'Integrated project delivery: An example of relational con-
    tracting'. *Lean Construction Journal*, 46–61, 2005.

NEC (2013) *How to use BIM with NEC 3 Contracts*, NEC, Glasgow, UK.

RIBA (2013) *Outline Plan of Work*. RIBA.

Strategic Forum (2007) *Profiting from Integration*, CIC, London, UK.

Thomsen, C., Darrington, J., Dunne, D. and Lichtig, W. (2009) *Managing Integrated Project
    Delivery*, Construction Management Association of America (CMAA), McLean, VA.

Williamson, O.E. (1979) 'Transaction-cost economics: The governance of contractual
    relations'. *Journal of Law and Economics*, 233–261, 1979.

Wolstenholme, A. (2009) 'Never waste a good crisis: a review of progress since Rethinking
    Construction and thoughts for our future'. Constructing Excellence. Online.
    Available HTTP: <http://constructingexcellence.org.uk/wp-content/uploads/2014/12/
    Wolstenholme_Report_Oct_20091.pdf>.

"WELL, THAT'S WHAT IT SAYS IN THE CHANGE ORDER..."

8 Contract administration

Wes Beaumont

Introduction

The construction industry is complex and vast, comprising some 280,000 self-governing, but interdependent organisations (BIS 2013). Changes in procurement and project delivery strategies, together with the move towards design and build, has only served to fuel this industry fragmentation. Figures 8.1 and 8.2 illustrate a comparison between a traditional and a design and build contractual structure reflecting the evolution of stakeholder relationships. Unbroken lines indicate contractual relationships and broken lines non-contractual relationships.

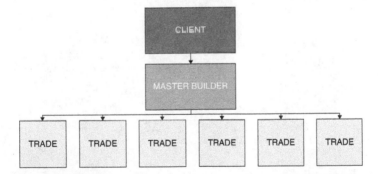

Figure 8.1 Traditional contractual structure.

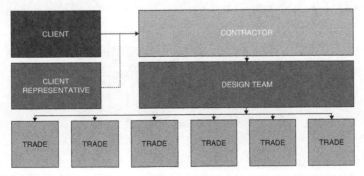

Figure 8.2 Design and build contractual structure.

The construction supply chain accounted for £124bn worth of *intermediate* consumption (BIS 2013) illustrating the sheer number of organisations participating in the sector, increasing the number of contractual and non-contractual relationships required. When the wider supply chain is considered, in contractual and non-contractual structures, the complexity increases significantly as Figure 8.3 (BSRIA 2005) illustrates. BIS (2013) identified that up to 70 subcontractors can be involved on a typical medium-sized project (circa £20m), further increasing the complexity of administering contracts.

A number of industry challenges originate due to project complexity, with large numbers of influencing actors and associated inter-dependencies. Complexity is defined as, 'consisting of many varied interrelated parts' (Baccarini, 1996). As projects have become more complex, the ability to manage and control time, cost and quality has become more challenging. Generally, the higher the project complexity, the greater the time and cost (Baccarini, 1996). Figure 8.4 illustrates the breakdown of project complexity. Virtually all projects are by definition multi-objective, with parties to the contract often having conflicting goals.

The Built Environment Industry Innovation Council (BEIIC, 2009) reported that the inability to correctly identify project goals results in poorly defined

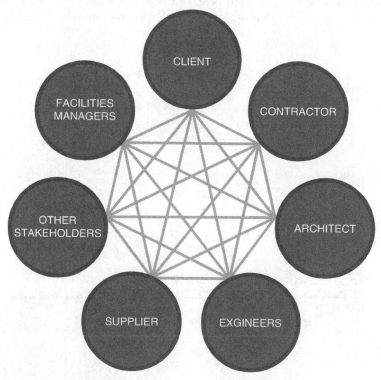

Figure 8.3 Traditional information exchange.
BSRIA (2013)

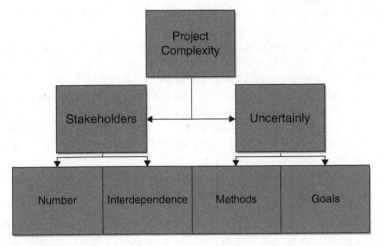

Figure 8.4 Project complexity.
Baccarini (1996)

scopes, instigating change, be that through variations or otherwise. The report, covering construction procurement generally, revealed the following:

- 52 per cent of respondents said the project they were involved in was inadequately scoped by the time the project was tendered.
- 64 per cent revealed that scoping inadequacies were identified too late.
- 60 per cent experienced cost and time overruns, by as much as 20 per cent, due to poorly defined scopes.

Poor client briefing and the lack of clarity around the employer's and contract requirements will inevitably lead to change. One of the key areas of concern for the Contract Administrator (CA) is the tracking of changes and the effect these have on cost and time. The CA needs to ensure transparency, fairness and the creation of equitable arrangements whilst administering the contract.

The deployment of BIM offers the opportunity for integrated management processes, along with improved project controls and validation of information. By identifying the employer's requirements early, including the Employer's Information Requirements (EIRs), and articulating these to the supply chain in a comprehensible way, together with processes that enable cooperation, the contract administration process may be improved, delivering benefits to all parties.

This chapter will, therefore, explore how contract administration is affected in projects that aspire to BIM Level 2 compliance. It will consider how this environment offers the opportunity to improve objectivity and proactive decision making to facilitate the effective delivery of a construction project.

Due to commercial sensitivities, it has not been possible to detail individual projects. However, all projects contributing to this chapter were administered by

Turner & Townsend with the exception of the Primary School project, which was managed by a local authority.

All of the projects were initiated during or after 2013, with values in excess of £8m across property sectors and infrastructure. Whilst the rationale for the deployment of BIM for each project differed, the creation of structured project information for use during the life cycle was a common theme throughout all projects.

The case studies discussed in this chapter look at the role of CA through the life cycle of a construction project, from interpreting the employer's and contract requirements, articulating these to the supply chain, and validating compliance through to handover.

Author biography

Wes Beaumont MCIOB is a chartered construction manager, and was a BIM Manager for Kier Group plc, leading the early implementation of BIM. He has supported a variety of Employers in process improvement and now works as a business analyst facilitating business improvement, primarily focused at the PMO level. He was a founding member of the Construction Industry Council BIM2050 group and co-authored the *Built Environment 2050* report. He has a first class honours BSc in construction management and an MSc with distinction in BIM and integrated design.

Company information

Turner & Townsend is an independent professional services company specialising in programme management, project management, cost management and consulting across the property, infrastructure and natural resources sectors. With 87 offices in 36 countries (2015), the practice draws on its global and industry experience to manage risk while maximising value and performance during the construction and operation of its clients' assets.

Service profile

As the RICS Guidance Note on Contract Administration published in 2011 indicates that, in its broadest sense, the role of CA has been around on building projects for centuries – although, interestingly, the term was not formally introduced until amendment 4 to the JCT 1980 form in 1987. The guide goes on to define the role as 'managing the contract between the employer and building contractor' (RICS, 2011).

In a traditional setting, project requirements are defined, refined and confirmed prior to construction commencing as illustrated in Figure 8.5. The RICS Guidance Note itself states that contract administration does not technically commence until a build contract is in place, however, in practice the responsibilities for contract administration will have commenced before the construction commences. Indeed, with the rise of design and build (D&B), the CA

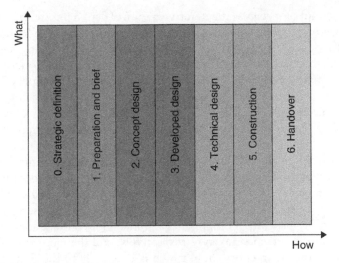

Figure 8.5 Traditional vs design and build development processes.

role commences in the design process and flows through both the design and construction phases, which in practice often progress in parallel. A key consideration in this scenario is how to control, manage and validate change requests for information (RFI), queries, valuations, payments and instructions. The figure below suggests that in a traditional setting the majority of the 'what' (the requirements) are determined prior to construction. In contrast, for a D&B scheme, the 'what' and the 'how' are often determined at the same time, which in some cases means that teams are determining *how* something is procured and constructed prior to the project being fully defined.

Scope of service

The RICS Guidance Note confirms that 'the CA is responsible for administering the terms of the building contract between the parties'. It goes on to say that 'the CA will act as the agent of the employer in some circumstances, but will be required to make impartial decisions' (RICS, 2011). Whilst the scope of services provided will vary depending on the nature of the project and the form of contract employed, the Guidance Note suggests that key CA tasks will typically include those illustrated in Appendix 1.

With the use of traditional contracting models, and hence processes, there is a greater propensity for inefficient information exchange, design iterations and longer review timeframes, all of which are compounded by the silo mentality in which each discipline performs its function in isolation from other stakeholders. Processes are generally based on stage gates, which close at specific stages, and this sequential process of generation, submission and review of information, can create reciprocal working, which adds little or no value (Eastman et al., 2011) and is exacerbated when changes are requested.

It is important to recognise the iterative nature of design information creation (Jamieson, 1997) and make allowances for concurrent working, which involves stakeholders 'cooperating', 'having equal authority' and 'working simultaneously or side by side'. However, this is something which traditional processes and contractual relationships largely prevent. The full extent of current 'collaborative processes' are limited, not concurrent and often only involve reactive problem solving, where each stakeholder understands only part of the overall problem (Eastman et al., 2011), often failing to appreciate the project and relevant issues holistically. Working in isolation, and towards deadlines at stage gates can prohibit prompt design changes, which may be further compromised by non-flexible arrangements often in place to determine the information required for design integration. Decision making and validation of changes can therefore add delays to a project, and due to time pressures, are often not considered holistically. This presents a problem for contract administration as project controls are undertaken retrospectively, and therefore are a function of reporting rather than controlling, which is inevitably to the detriment of a project.

The impact of BIM

As BIM is underpinned by the sharing of information, the issues of liability and intellectual property must be addressed. Whilst the CIC BIM Protocol tackles the issue of information reuse and licence acquisition (see Chapter 7), there remains the need for data and information to be validated for accuracy and completeness, particularly if that information is to be used to administer the contract. There also remains the distinct need to ensure that information created meets the specific needs of a process, i.e. for identifying a target cost at the appropriate time or stage.

The use of BIM allows for better sharing and singularity of information, which aids project delivery, enabling the project team to have a control process to deal with the changes as they occur. BIM requires a clear identification of requirements, supported by a robust process for articulating how those requirements are to be met and explicit validation procedures for checking deliverables at key stages within the project information life cycle. With such processes in place, the CA has the opportunity to verify where the project is failing to adhere to the contract requirements, and if appropriate, deploy processes to provide remedial or corrective actions. The use of BIM can underpin this process and provide the necessary framework for delivery.

Appendix 1 illustrates how the deployment of BIM can affect the CA role. It should be noted that not all functions will be impacted, albeit that a BIM environment will provide more accurate and precise information to administer the contract.

How BIM could be used

For BIM to be a success on any project requires a robust BIM Execution Plan (BEP), which responds directly to the employer's requirements. This implies a fundamental requirement to ensure clarity of objectives, responsibilities, deliverables and validation processes. Understanding what is to be measured throughout a project, provides a greater focus on what information is required. Plain Language Questions (PLQs) pose the question to the employer of 'why am I asking for this?', which is the requirement to state why certain information or data is required so that the project stakeholders can ensure that all requirements have a deliverable, and an objective, in meeting the business case through improved administration of the contract.

The deployment of BIM supports contract administration in that all project information should be located in the common data environment (CDE), Figure 8.6, and be of an appropriate format to support the information requirements set out in the EIRs and the BEP. This enables information to be *pulled* from the CDE in contrast to the traditional method of information being *pushed* by individuals. This can significantly reduce the risk of stakeholders using incorrect information or being misinformed, thereby promoting collaborative working.

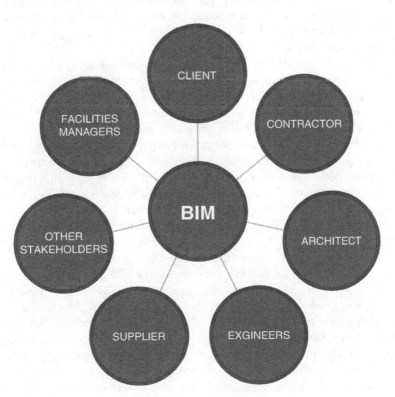

Figure 8.6 BIM creates a common data environment.

The CDE enables the project team to avoid the risk of people working with outdated information. Instructions can be issued electronically with many BIM solution platforms providing an integral task system to provide a contemporaneous communication trail as well as a target date for instructions to be completed. Tasks can be deemed not to be completed until the works have been confirmed and the required data entered into the system.

As the management of change forms a significant part of the CA role, it is essential that project changes are effectively communicated to all relevant parties (Cooper et al., 2005). The use of data visualisation[1] allows changes to be highlighted and provides an auditable trail of design iterations as well as on-site production and installation for increased levels of responsibility and liability control. The use of model authoring software and model components with parametric properties can mean that when changes are introduced all associated data can be updated simultaneously.

Changes can be monitored by adopting a CDE in accordance with PAS1192-2:2013, which is intended to promote a collaborative environment through which site teams can share the same common information pool which gives all team members the ability to identify, communicate and address any issues likely to result in defects and/or variations, thereby reducing or even eliminating the need for on-site design development. It is important to ensure that permitted use and access rights are maintained to avoid the use of information that has yet to be validated and signed off. BIM solutions will typically synchronise information across all relevant project stakeholders.

The use of cloud-based-BIM compatible mobile technology (Figure 8.7) enables the site team to communicate with other project stakeholders. Changes can be tracked by using information models and overlaying them to identify changes. A key benefit here is the ability to identify if a change is due to a client variation or whether it is simply design development or to resolve an error.

The quality of information is of paramount importance when administrating a contract. If the quality and consistency of information produced through the life cycle is poor then there is a distinct risk of decision making becoming inconsistent and poor judgements being made. Simply put, the decision making information has to be accurate and up to date to ensure that the correct decision can be extrapolated. By quality assuring the structured information created during the BIM process, using technology and embedded rule sets, information can be assured akin to that of a 'digital information clerk of works'. Rule sets can be used to check information validity, for example, Disability Discrimination Act 1995 (DDA) compliance can be checked in this way i.e. checking the width of a door to ensure a wheelchair can pass. This role has been defined by the CIC BIM Protocol as part of information management, a role which could be undertaken by the QS (see Chapter 6). It is therefore increasingly likely that as more clients procure both physical and digital assets, information management will play a progressively more significant role in supporting contract administration.

The move from analogue processes, which are often subjective and inefficient, to digital processes, which are objective, efficient and precise, will improve

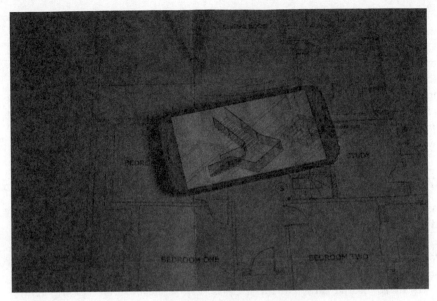

Figure 8.7 The use of mobile BIM technology.

contract administration and assist in providing proactive analytics to forecast, predict and improve decision making. The use of data derived from the information models, which is validated against the employer's requirements, will provide a clear and objective comparison between requirements and deliverables.

By adopting digital processes, and ensuring that data is consistently structured, the CA can utilise data visualisation techniques to report the current status of a project in contrast to creating the more traditional manually prepared and often subjective reports. For example, as costs are progressively developed through the pre-construction phase, the effect of changes on the design and/or specification can be tracked, assisting with identifying the cause and effect of such changes.

Adopting this approach removes the need for subjective decision making and facilitates more objective decisions improving transparency. One obvious example is the ability to track progress on site using the planned programme against actual progress. Additionally, using data visualisation to model and assess impact helps to identify and mitigate potential problems before they occur (see also Chapter 3, Risk and risk management).

Case study details

The following is based on the experience of delivering CA services across a number of projects delivered using BIM and includes an overview of a BIM Level 2-compliant project from inception to handover and operation. It includes references to a number of projects where Turner & Townsend (T&T) were engaged as CA and provides an insight into how BIM supports contract administration. It commences with a more in-depth exploration of the employer and contract requirements, forming the EIRs. Following this a BEP is created, essentially governing how the contract will be delivered to meet the EIRs.

The Primary School is an example of a BIM early adopter project for a local authority. The local authority (LA) are relatively unusual in that all design is undertaken in-house. The LA decided to run two similar projects in parallel. one utilising traditional 2D design and construction techniques, and the other in compliance with BIM Level 2. The Primary School was an £8m development situated on a new mixed use and residential site. The new school provided 60 reception places as well as playground and sports facilities.

Why BIM was used

As projects progress, the quantity of information required to meet the requirements of the employer and the contract increases, along with its relative importance and detail. This necessitates robust information management and validation to ensure that during construction the documentation is accurate, precise and relevant, thereby reducing errors, RFIs, risk and variations.

As already noted earlier in this chapter, whilst the rationale for the deployment of BIM varied for each project, the creation of structured information for use over the project life cycle was a common theme throughout all projects.

How BIM was used

In the case of all projects referenced, the level of detail (LOD) required at each stage was identified and articulated via the model progression delivery table (MPDT). This replaced traditional information requests and the provision of static drawings and schedules. Together with the introduction of PLQs, the contract requirements were such that it was explicitly clear exactly what was required at each stage to meet the needs of the contract. This helped to avoid scope creep and the potential for variations later during the contract.

The LA deployed the use of BIM during the Primary School project and compared the outputs during the construction phase against a traditional project; the differences are illustrated in Table 8.1 below.

The T&T approach to CA is predicated on the information validation model illustrated in Figure 8.8, and it is based on five key checks on five areas relating to a project.

A series of checks are applied on extracted data and the model itself using in-house authored tools across five key areas; commercial, level of development, quality, design management and model data. Generally the information provided at key stages is measured against data requested in the EIRs and as reflected in the BEP. This validation model is designed to provide answers to the following:

- Commercial metrics – does the extracted data benchmark favourably against established client key performance indicators (KPIs) such as GIFA or

Table 8.1 Comparison between traditional and BIM outputs.

Project	Traditional Primary School	BIM Primary School
Information	2D coordinated drawings	3D federated model
Output	2D drawings and specification	2D drawings, specification and 3D models
Contractor	Early engagement and review of drawings	Early engagement and review of drawings and models
Programme	Developed from information received	Developed from information received and models used to inform decisions
M&E strategy	Services to pass through holes in beams above ceiling level. Complex drawings required to illustrate strategy	Services to pass through a 350mm void above ceiling level (spatial coordination). All services runs modelled in 3D prior to tender
Tender outcome	Value engineering required to reduce	Within available funding
Target net rate	Not achieved	Achieved

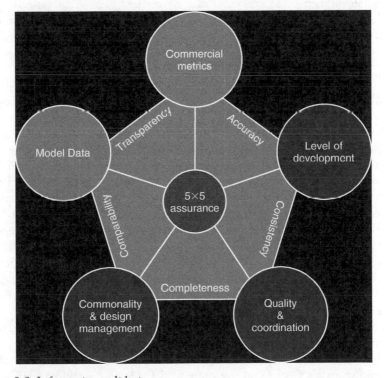

Figure 8.8 Information validation.

occupancy levels? What are the changes that have occurred since the last iteration of the project and what effect does that have on cost?
• Level of Development – does the model data support the uses that are required of it at the right stages? Is the information provided 'fit for purpose'

and can the site teams use it to construct the project according to the contract requirements?

- Model quality and coordination – has the authoring been consistent and coordination applied? Can the design be signed off at the appropriate stage?
- Commonality of components and design management – have efficiencies been driven into the process at every stage and are the right parameters embedded? Do the specified components meet the contract requirements?
- Model data – is it complete and correctly structured and are parameters correct? Does the data enable reporting on cost and programme?

This process is designed to check that information is fit for purpose and meets the contractual requirements. It encourages more transparent governance as decisions can be made more objectively with the availability of data created within the supply chain and assessed against pre-agreed contractual obligations. T&T have also experienced using the information model for commercial metrics, such as earned value analysis and change control, which assists in controlling cost as changes, amendments and omissions may be tracked between iterations, enabling the effect of changes to be quantified by the CA as work progresses through the relevant work stages.

Tools used

A BIM solution designed to support all phases of the asset delivery life cycle including asset management was deployed.

A cloud-based-BIM compatible mobile technology solution was identified as a solution which would provide the desired visibility of the programme and facilitate collaboration between all members of the project team. It provided the primary means of communication between the supply chain and the client, ensuring that all site team members had the latest iteration of models and associated information, that they were aware of any changes to the programme, whilst at the same time providing simple platform to confirm quality and progress against key metrics.

Contract administration was improved by the use of these tools and through the early receipt of the latest information by all parties. Timely information made contract administration far simpler and concurrent to the construction process, allowing issues to be identified early, which in turn allowed the CA to plan accordingly and, where possible, to advise the Project Manager (PM) to take appropriate action to mitigate impact on the project.

T&T also tracked construction progress using the planned construction simulation and verifying progress on site using field BIM tools, enabling the CA to ascertain whether progress was on schedule, and if not, the reason for any delay. A similar regime could be used with cost by accessing the dynamic model; cost managers are able to identify the root cause of any changes and to ascertain the rationale for the change. This information can then be presented objectively to enable fair and transparent decisions.

20%	30%	**Executive summary**
Level of Development	Model Data	**Risk**

Lorem ipsum dolor sit amet, consectetur adipiscing elit. Cras posuere ante sed elit ullamcorper mattis. Fusce rutrum ligula ut laoreet lacinia. Morbi vel metus fermentum, tincidunt sapien eget, porttitor lorem. Nam vulputate odio magna, ut vehicula sapien eleifend sit amet. Donec dictum ante malesuada metus posuere, nec scelerisque dui blandit. Morbi arcu enim, ornare a fermentum et, scelerisque eu odio. Aliquam et augue ut nisi eleifend imperdiet. Nunc et mi ligula. Cras augue arcu, fringilla vitae cursus ac, viverra sed enim. Pellentesque faucibus varius massa et venenatis. Suspendisse scelerisque justo tincidunt dolor mollis semper. Ut pretium et lacus nec congue. Duis porttitor accumsan urna.

Drivers

Class aptent taciti sociosqu ad litora torquent per conubia nostra, per inceptos himenaeos. Sed ullamcorper est non rutrum rhoncus. Nullam et ullamcorper erat, vel malesuada massa. Cras vehicula nisi non nulla rhoncus facilisis. Aenean ultricies laoreet luctus. Aliquam vel velit ornare, semper arcu vitae, gravida orci. Etiam vitae dui au neque mattis hendrerit. nam auctor ornare tellus vitae egestas. Praesent imperdiet nisi nec risus vestibulum condimentum. Sed porta justo eget justo commodo, eu semper urna ultrices. Maecenas non tincidunt est, vitae pellentesque eros. Proin nunc sem, ornare id pulvinar a, vestibulum sit amet nibh.

Opportunity

Sed ut ornare elit. Sed blandit, diam tempor tempor tempor, nisl sem iaculis odio, scelerisque congue ipsum elit fermentum neque. Etiam sed consectetur arcu, eget porta augue. Nunc dignissim diam nibh, ac placerat tortor faucibus id. Suspendisse potenti. Integer tincidunt pellentesque nibh, non pellentesque dui suscipit vel. Praesent vehicula auctor ipsum at viverra. Morbi posuere blandit ultrices. Pellentesque habitant morbi tristique senectus et netus et malesuada fames ac turpis egestas. Integer id iaculis lectur. Curabitur venenatis purus ut suscipit volutpat. Donec molestie erat a augue lobortis, sit amet sollicitudin nisi rutrum.

Figure 8.9 Reporting dashboard.

Model Data Summary

Resp Party	COBie she..	Rule	Checked Componen...	Issue	Sum of Compliance
ARC	Facility	Project Should Have Area Measurement	£90	YES	47%
		Site Description Should Exist	343	YES	47%
		Site Should Have Description	376	YES	47%
	Floor	Floor Category Value From Picklist	114	Yes	43%
		Floor Description if AECOsim Models Exist	339	YES	51%
		Floors Should Have Elevation	206	YES	44%
		Floors Should Have Height	216	YES	70%
	Space	Space Category - Uniclass Classification Required	218	YES	66%
		Space Name Should Exist	256	YES	34%
		Space Should Have Gross Area (Gross Floor Area)	967	YES	44%
	Type	Type Category - Uniclass Classification Required (improved)	258	YES	68%
		Type Should Have Name	130	YES	39%
		Type Should Have Nominal Height	124	YES	77%
		Type Should Have Nominal Length	259	YES	45%
		Type Should Have Nominal Width	271	YES	38%
	Zone	Check zoneClassificationCode of space is consistent	166	YES	73%
		Zone Category - Classification from Picklist.	165	YES	23%
		Zone Category - ZoneClassificationCode Exists	281	YES	71%
		Zone Name Should Be Unique	111	YES	41%
		Zone Should Have Name	286	YES	52%

0% 10% 20% 30% 40% 50% 60% 70% 80%

0% ▬▬▬▬▬▬▬▬▬ 100%

Resp Party
- ARC
- CON
- MEP
- STR

Figure 8.10 Interactive dashboard example.

Analysing large data sets, created and stored during a typical project, supports proactive decision making based on trends and analytics through the use of dashboard technology (Figures 8.9 and 8.10). Interactive dashboards enable the CA to manage by exception i.e. monitor at a high level and then drill down into areas of concern, finding the root cause and taking appropriate action. The reports are dynamic, which allows efficient tracking of progress.

Utilising data visualisation for reporting purposes enables the project team to make decisions quickly, as cause and effect may be identified more easily. In the same way that visualisation from BIM enables all stakeholders to understand the project, data presented in the form of a dashboard similar to the example provided above enables an understanding of any issues likely to have a time and/ or cost impact on the project and therefore the role of CA.

Issues/benefits

There are some pitfalls to utilising BIM and data management to support the contract administration process. One key risk is making use of data that has not been structured correctly and/or validated. By trusting this information and using it to make key decisions could lead to wrong decisions being made which will be to the real detriment of the project.

In the case of the Primary School project, the client and the contractor worked collaboratively to develop a comprehensive and cohesive construction programme. In order to gain the most out of BIM, the client recognised that they needed to invest the same effort in the construction process as they did for the design process. The LA were aware that traditional, almost retrospective, construction management processes would not provide them with the desired visibility of the programme to support decision making.

To summarise, Table 8.2 highlights the differences between a traditional project delivery strategy and that used for the Primary School, where BIM was deployed during the construction stage.

Typically, on a project of this size the client would expect to allocate two members of staff. With the deployment of BIM, only one member of staff was required on a part-time basis to manage delivery of the project, which significantly reduced the resource overhead. The BIM process additionally provided a structure which defined and established key information delivery milestones. These were easy to follow and, perhaps more importantly, the data requirements were limited to those defined in the EIRs to meet the client's needs rather than attempting to capture everything.

One major benefit of utilising BIM and data management in construction, as already stated, is the opportunity to make objective decisions. T&T now adopt data visualisation to report on contract administration activities, utilising the data derived during the validation stage as discussed earlier. The use of data visualisation and analytics provides an intuitive and accurate representation of

Table 8.2 Benefits of traditional vs BIM projects.

Project	Traditional Primary School	BIM Primary School
Information	2D coordinated drawings	3D federated model
Progress	Completed behind schedule	Completed on time
Information requests	High number of RFIs	Low number of RFIs
Services coordination	Additional resources required to resolve multiple issues	No issues reported
Management Resources required by the client to deliver the project	2 individuals full-time	1 individual part-time

the current status of a project including reporting on the costs and programme, as well as tracking the status of instructions. During the construction stage the reports include areas such as status on installation of components, snagging and quality management and commissioning.

The enhanced coordination process utilised on a BIM project removes ambiguity at the design stage. Early access to coordinated model information and site programme meetings using the associated 3D model views, provides early opportunities for site operatives to see how they would construct or install their systems prior to works commencing and for any identified issues to be resolved before they impact on the project.

The introduction of BIM, therefore, affords greater clarity when deciding if a change is due to design development or a variation to the works. Furthermore, as models become *the* method of controlling design, and as more objects and components are modelled correctly, the model can be utilised for measurement and/or re-measurement. This ensures that the CA can value any changes to enable fair and appropriate payment to be made to the contractor and the final account to be assessed with greater accuracy.

In one of the case studies, the Site Manager kept a record of issues that would have occurred had he not been using a BIM-compatible mobile technology. The potential cost of rework was subsequently identified and demonstrated £48k of confirmed savings with zero rework.

A single reporting medium also helps ensure that all stakeholders are aware of current project status. Having a shared database presented visually improves accessibility, understanding and objectivity.

As BIM can afford timely information, the role of the CA can be made far simpler and concurrent to the construction process. The main benefits being that issues may be identified early, which allows the CA and the PM to plan accordingly and where possible avoid particular problems, minimising on-site development.

The process devised by T&T provides confidence in the information that underpins the contract administration process. It reduces risk by increasing the availability of information, along with the usability of that information. At key milestones identified in the EIRs and BEP, and in line with work stages, information is coordinated, validated, assured and verified. Identifying and correcting discrepancies in the data derived from BIM removes the opportunity for errors in scheduling, quantification and analytics. As previously discussed, the use of 'rule-based' checking using proprietary software helps ensure that all information required to support decision making is inherent, accurate and precise. The results of the validation process are then presented using data visualisation. Ultimately, however, the validation process provides confidence in the information being produced for use during construction.

During the Primary School project, RFIs were provided to the design team, digitally supported by photographs and 3D views which were navigable (rather than pictures of 3D views). More importantly, this provided the client with the information about any issues early allowing time to evaluate the problem

Table 8.3 How BIM alleviates common CA pitfalls.

Common pitfall	How BIM can assist
Excluding key stakeholders leading to unnecessary starts and stops and unclear user requirements and scope creep	The EIR'S and initial COBie drop at digital Plan of Work (dPOW) stage 1 creates more explicit user requirements by encouraging a more detailed project scoping process and greater definition. These requirements are articulated to the supply chain resulting in a BEP
Poorly defined scope of works that does not reflect the agreement between parties	Creation of a single responsibility matrix – the 'Model production and delivery table', defined in PAS 1192-2:2013, which explicitly identifies information requirements and responsibilities through the project life cycle
Over engineering processes leading to difficulty in use and problems with reporting	Validation – gateway points with COBie drops and data visualisation reporting
Poorly defining objectives of the employer	Inclusion of PLQ's which indicate an objective at each defined work stage, supporting the client objectives and the validation of contract deliverables.
Change orders occurring due to poor information and lack of coordination	Explicit identification of information needs and coordination of design and associated information in a timely manner with the regularly sharing of all project information. This additional clarity helps avoid disputes and rework. The variation process is made far simpler and transparent.
Information not delivered at the correct time	Information needs are defined in the EIR and through exchange points in the BEP. Deliverables are validated for compliance using rule-based software e.g. clash detection and avoidance
Information created in bespoke formats difficult to validate	The BEP defines exchange formats to ensure interoperability to enable the CA to interrogate information.
Information stored in various locations	The common data environment (CDE) stores information in a common and secure location.
Opportunism to benefit from uncertainty as to project scope and final price	Models contain all the attributes to scope and price to be defined. Models are shared regular to enable a robust audit trail to be created.

and define a solution, rather than being forced to 'work around a subcontractor solution', be exposed to variations, or have to fund any rework. This process also reduced the need and frequency for the architect to visit the site by circa 60 per cent. The deployment of BIM and the increased clarity of requirements it enabled also resulted in a 70 per cent reduction in RFIs.

Table 8.3 summarises some of the common pitfalls with traditional contract administration and the opportunities and benefits that the use of BIM presents.

The key benefits of BIM for the role of CA may, therefore, be summarised as follows:

- Early and more accurate definition of the employer and contract requirements, which are included within the EIRs. Ensuring that specific information requirements and stage deliverables are defined as early as possible.
- Identify and articulate roles and responsibilities, including strategic obligations relating to information creation and reuse, reducing duplicate effort and ensuring that information is provided in an appropriate format for compliance validation.
- Make it simpler to evaluate and identify changes earlier along with their cause and effect.
- Serve as a communication aid between all parties to the contract, with changes updated in a timely manner to ensure that all parties are aware of the current status of the project. This is particularly important during the construction phase. Changes are validated and assured prior to being signed off by the employer, and implemented.
- Assist in the validation process to ensure that the employer's requirements are achieved.
- Report using data visualisation, reporting against the EIRs, enabling objective decision making and improving transparency.
- Changes to record drawings were minimal which is unusual (this was seen as a key benefit).
- Variations recorded were due to design decisions and not corrective action.

Lessons learned

Some of the lessons learned during the case studies include:

- It is critical to identify objectives early and to ensure that design progresses collaboratively, and that project controls and validation procedures are understood by the project team.
- The requirement to develop information models early requires more time initially. However, the benefits are provided during construction as noted during the Primary School project.
- BIM offers a change management environment, but this does require buy-in from all stakeholders to be effective.

Summary/commentary

The use of BIM and data management greatly improves the commercial management of a construction project through better governance, improved information, enhanced transparency and greater objectivity. Storing structured, consistent and accurate information, which has been validated and assured, in a shared common data environment (CDE) reduces wasteful processes and helps to ensure that all stakeholders are working to the same information, often enabling multiple uses from the same information. Stakeholders are abreast of the same information and are acutely aware of the project's current status, again creating more equitable and transparent processes. All of this combined provides a far greater level of objectivity and control to both support and enhance the CA role during the project delivery life cycle.

It is perhaps important to stress here that the CA role will still be governed by the form of contract in place in terms of the mechanisms for executing the required function and role, i.e. administering instructions, valuations and payments or other notices required, in compliance with the conditions of contract. This means that the underlying 'administration' will remain largely unchanged, and that it will be in the management and execution of the CA role that BIM will have the greatest impact (see Appendix 1).

References

BIS (2013) *UK Construction – An Analysis of the Sector*, Department of Business Innovation and Skills, London, UK.

BSI (2013) *PAS1192-2 Specification for information management for the capital/delivery phase of construction projects using building information modelling*, British Standards Institute, London.

BSRIA (2005) *Marketing the Integrated Team*, BSRIA, Bracknell, UK.

Cooper, R., Aouad, G., Lee, A., Wu, S., Fleming, A. and Kagioglou, M. (2005) *Process Management in Design and Construction*, Blackwell, Oxford.

Eastman, C., Teicholz, P., Sacks, R. and Liston, K. (2011) *BIM Handbook*, Wiley and Sons, Hoboken, NJ.

Jamieson, I. (1997) 'Development of a Construction Process Protocol to promote a Concurrent Engineering Environment within the Irish construction industry', *First International Conference on Concurrent Engineering in Construction*, 4–5 July 1997.

RICS (2011) *Contract Administration: RICS Guidance Note, first edition*, RICS, Coventry.

Further Reading

Assaf, S., Mohammed, A. and Muhammad, A. (1995) 'Causes of Delay in Large Building Construction Projects', *Journal of Management and Engineering*, 11(2): 45–50.

Baccarini, D. (1996) 'The concept of project complexity: a review', *International Journal of Project Management*, 14(4): 201–204.

Baldwin, A., Shen, G. and Brandon, A. (2009) *Collaborative Construction Information Management* Spon Press, Oxon.

Banwell, H. (1964) *Report of the committee on the placing and management of contracts for building and civil engineering work*, HMSO, London.

BEIIC (2009) *Built Environment Procurement Practice: Impediments to Innovation and Opportunities for Changes*, Built Environment Industry Innovation Council, Australia.

BSI (2007) *BS1192:2007 Collaborative production of architectural, engineering and construction information code of practice*, British Standards Institute, London.

Constructing Excellence (2004) *Lean Construction*, Constructing Excellence in the Built Environment, London.

Cooper, R., Kagioglou, M., Auoad, G., Hinks, J., Sexton, M. and Sheath, D. (1998) *The Development of a Generic Design and Construction Process* in Proc European Conference Product Data Technology Building Research Establishment, Watford.

Appendix 1: Contract administration and the impact of BIM

Contract Administration Function	Traditional Activity	Impact of BIM
General administration, including managing the provision of information, finance and supervision	Interim valuations and certifying payments.	Use of the model(s) can assist in the valuation of completed works & re-measurement of work.
	If quality and standards have been identified as the CA's role then they shall be to the CA's reasonable satisfaction.	Implementation of a CDE in accordance with PAS 1192-2:2013 to improve information exchange and visibility.
	Maintain accurate and accessible records. Archive on completion. Recording actions and events that take place during the project relevant to the performance of the contractual obligations of all parties.	The management of many drawings, paper based documents and then design development can be time consuming, complex and full of pitfalls. This can lead to miscommunication, misinterpretation and increased levels of risk. The use a CDE from the outset in accordance with PAS 1192-2:2013 improves understanding, change management and information exchange.
		Management of information exchange is critical and a CDE can remove problems of information handling and fragmentation in communication methods such as emails and the "who has what?" scenario associated with traditional paper based modes of communication. These traditional methods can create adversarial conflict and lead to poor productivity, whereas a CDE can provide an effective communication infrastructure.
Records, inspection and general correspondence	Regular site inspections may include some or all of the following (subject to the CA's agreed roles and responsibilities): • Quality of workmanship related to the contract documents • Review of progress	Mobile technology may be used to upload and share snagging schedules/photos etc. Compliance with PAS1192-2:2013 can assist with managing change control by comparing iterations of the model(s) to identify changes, amendments or omissions. The CDE document structure and uploading process should be documented in the BIM Execution Plan (BEP) and places a contractual obligation to facilitate the process.

Contract Administration Function	Traditional Activity	Impact of BIM
	• Check material being used • Check work conforms with specifications and drawings • Records of any measurement of work General overview of health & safety arrangements on site.	With the use of mobile technology, site personnel can take, store and share photographs of progress providing contemporaneous records of constructed elements. For example, steel reinforcement within pad foundations.
Site inspections, quality, progress and health and safety	Agree a programme of design team meetings at project inception and progress meetings (set agendas) at the commencement of a contract. Confirm the status of the necessary statutory comments and obligations – e.g planning.	Progress can be monitored using a schedule based simulation model to compare actual progress versus planned progress. The principles of the BIM methodology encourages collaboration and increased levels of design development during the pre-construction phase ie. dPoW work stages 1-4.
Meetings, general and statutory matters	Ensuring compliance rather than authorising including witnessing /administrating testing. Identify the obligations of all parties – report on the progress of parties discharging their obligations. Reporting on the financial position of the project on a regular basis.	The use of shared information further facilitates collaboration with all project information being open and transparent between project stakeholders.
Reporting to the client/employer, dealing with claims and variations	Issue instructions to change work. Manage omissions from the project.	As models are shared frequently and developed in line with a defined LOD they can be assessed for compliance against the EIR'S. Variations may be assessed to determine the impact.

Contract Administration Function	Traditional Activity	Impact of BIM
Client/employer instructions, changes and cost savings	Advising on any risk of delay to the contractual completion date. Monitoring progress of the works.	Ability to better monitor design development improving understanding and assist with change management. Making more informed decisions at earlier stages can reduce the level of redesign and rework later in the process and in turn the cost associated with change orders avoiding disputes.
Programming and impact of changes.	Administering the process in accordance with the contract rather than authoring.	Use of schedule based BIM to assess the contractor's intent associated with construction sequencing and coordination.
	Advise and report the cost implications of variations.	BIM integrated with cost and scheduling can provide more accurate cash flow forecasting. This can also assist in funding and drawdown planning and assessment of progress against programme.
Contract instructions/ variations	Issue the correct notices and certificates aligned with the contract requirements.	Use of the BIM and cost based compatible software to monitor design development and change.
	Assess entitlement for extension of time, the requirement for early or partial possessions and completion of all works.	Use of schedule based BIM software to monitor time implications of instructions or variations.
Contract completion date, extension of time (EoT), partial possession and practical completion	Assess and approve/reject.	Use of schedule based BIM software to monitor and manage the programme. Contractor to keep the programme up to date and publish to the CA.
Loss &/or Expense	Maintain and report.	Use of BIM integrated with cost and scheduling to monitor and assess claims for loss and/or expense.
Adjusted contract sum/ final account	Maintain and report.	Use of the BIM and cost based software to maintain and report the likely adjusted contract sum (ACS).

Davies, A. (2009) *From Iconic Design to Lost Luggage: Innovation at Heathrow Terminal 5*, Imperial College, London.

Davison, B. and Sebastian, R. (2006) *The relationship between contract administration problems and contract types*, International public procurement conference proceeding, 21–23 September 2006.

Deutsch, R. (2011) *BIM and Integrated Design*, Wiley and Sons, Hoboken, NJ.

Edmonson, R. (2002) 'Integrated supply chains and building services suppliers', CIBSE National Conference 2002.

Egan, J. (1998) *Rethinking Construction: report from the Construction Task Force*, Department of the Environment, Transport and Regions, London.

Emmerson, H. (1962) *Survey of problems before the construction industries*, HMSO, London.

Fallon, K. and Palmer, M. (2007) *General Buildings Information Handover Guide: Principles, Methodology and Case Studies. An Industry Sector Guide of the Information Handover Guide Series*, Department of Commerce, United States of America.

Fisher, N. and Morledge, R. (2002) 'Supply chain management' in Kelly J, Morledge R, and Wilkinson, S (eds.) *Best Value in Construction*, Blackwell, London.

Harding, C. (2010) 'Integrated design and construction: divided we fall', *Building Magazine*, 29 October 2010.

HM Gov (2010) *Low carbon construction innovation and growth team final report*, HMSO, London.

Johnson, G., Scholes, K. and Whittington, R. (2005) *Exploring Corporate Strategy*, Pearson, Essex.

Kagioglo, M., Cooper, R. and Aouad, G. (1999) 'Re-engineering the UK construction industry: the process protocol', *Proceedings of the Second International Conference on Construction Process Re-Engineering*, University of New South Wales, Sydney.

Kelly, J., Morledge, R. and Wilkinson, S. (2002) *Best Value in Construction*, Blackwell, Oxford.

Khanzode, A., Fischer, M., Reed, D. and Ballard, G. (2006) *A Guide to Applying the Principles of Virtual Design & Construction (VDC) to the Lean Project Delivery Process*, Centre for Integrated Facility Engineering, Stanford University, CA.

Koskela, L. (1992) *Application of the new production philosophy to construction*, Technical Report 72, Centre for Integrated Facility Engineering, Stanford University, CA.

Kymmell, W. (2008) *Building Information Modelling: Planning and Managing Construction Projects with 4D CAD and Simulations*, McGraw Hill, New York.

Latham, M. (1994) *Constructing the team: final report of the government/industry review of procurement and contractual arrangements in the UK construction industry*, HMSO, London.

McKeeken, R. (2010) 'Here's to a more open relationship', *Building Magazine*, 10 December 2010.

Mosey, D. (2010) 'A decade on from PPC2000: carry on partnering', *Building Magazine*, 15 October 2010.

Neilsen (2007) *Profiting from Integration*, Constructing Industry Council, UK.

Park, B. and Meier, R. (2007) 'Reality-based construction project management: a constraint-based 4D simulation environment', *Journal of Industrial Technology*, 1 (23).

Sarshar, M. (2006) *Sharing Good Practice across Construction Organisations: the Search Continues*, Liverpool John Moores University, Liverpool

Scarbrough, H. (2000) *Investigating Knowledge Management*, Chartered Institute of Personnel and Development, UK.

Selectica (2013) '10 contract management implementation pitfalls' Online.

Available HTTP: <http://www.slideshare.net/Selectica/10-contract-management-implementation-pitfalls>, accessed October 2014.

Shelbourn, M., Bouchlaghem, M., Koseoglu, O. and Erdogan, B. (2005) Collaborative Working and its Effect on the AEC Organisation in *Proceedings of the 2005 ASCE International Conference on Computing in Civil Engineering*, Cancun, Mexico.

Shen, G., Brandon, P. and Baldwin, A. (2009) *Collaborative Construction Information Management*, Taylor and Francis, Abingdon, Oxon.

Shourangiz, E., Mohamad, M., Hassanabadi, M., Banihashemi, S., Bakhtiari, M. and Torabi, M. (2011) *Flexibility of BIM towards Design Change* International Conference on Construction and Project Management, Singapore.

Staub-French, S. and Khanzode, A. (2006) '3D and 4D modelling for design and construction: issues and lessons learned', *ITcon*, Vol. 12.

Sun, M., Anumba, C. and Sexton, M. (2004) *Managing Changes in Construction Projects*, UWE, Bristol.

Udom, K. (2012) *BIM: mapping out the legal issues*, The NBS, RIBA Enterprises.

9 Performance measurement and management

Rob Garvey

Introduction

Performance management should be an integral aspect of managing projects and organisations, aimed at ensuring regulatory compliance monitoring, controlling progress and supporting continuous improvement. Effective performance management is predicated on collating appropriate data to provide the information that will inform managers on performance. From the number of government-sponsored reports, it would be reasonable to conclude that the construction industry has a history of underperforming. Indeed, Sir John Egan's seminal report *Rethinking Construction* (1998 p.4) suggested that too many of the industry's clients were dissatisfied with the industry's performance in terms of time, cost and quality. Subsequently, *Rethinking Construction* provided the impetus for the industry to consider how performance could be more effectively managed and improved. This chapter will focus on the evolution of performance management in construction and consider the impact of Building Information Modelling (BIM). The emphasis from a BIM perspective is on information management, exploring how the data generated from digital environments created by BIM can facilitate more effective performance management, and how BIM can have a significant role in the evolution of construction performance management.

From a quantity surveying perspective, measurement is a core competence alongside good information management and the management of performance. Indeed, the Building Cost Information Service (BCIS) established by the Royal Institution of Chartered Surveyors (RICS) in 1961 sought to collate and analyse cost information and subsequently provide benchmarking facilities to measure and assess performance. The services of the quantity surveyor (QS) are based on providing information that facilitates effective decision making, which can be enhanced by the introduction of BIM. However, Barker (2011) highlights that in addition to traditional quantity surveying skills, BIM demands new skills that 'cross the traditional boundaries', such as data management, 3D modelling and 'technical skills to resolve difficulties at the interface of processes'. This chapter looks at how and where these skills might be applied to develop services to support performance measurement. It is recognised that performance measurement is likely to form a part of a wider service delivered by the QS, be that project management, contract administration or cost management.

Author biography

Rob Garvey is a Senior Lecturer at the University of Westminster teaching on undergraduate and postgraduate courses in the Department of Property and Construction within the Faculty of Architecture and the Built Environment. A Chartered Quantity Surveyor with more than 20 year's industry experience working with Mace and other Tier 1 suppliers on projects, such as London Heathrow Terminal 5, Rob specialises in all commercial aspects of both project and organisational management. Rob is Academic Partner on the Government Trial Projects for procurement; monitoring a project approach to collaboration and integration.

Organisation information

The University of Westminster has more than 20,000 students from over 150 nations on practice-based courses, which are independently rated as excellent, many with international recognition. Its distinguished 175-year history has meant it leads the way in many areas of research, particularly politics, media, art and design, architecture and biomedical sciences, and its position in the city of London allows it to continue to build on close connections with leading figures and organisations in these areas as well as in the worlds of business, information technology, politics and law.

What is performance measurement and management?

Performance measurement has become a significant strand of a manager's toolkit. Measures of performance are regularly used to comment on some aspect of business or public sector performance in newspapers and the media, with the latest crime figures, health service waiting times, poor customer service or other similar reports generated by performance statistics becoming part of everyday life. However, performance can be open to interpretation and this section clarifies the meaning of performance to provide some understanding of why performance can mean different things to different people at different times.

If performance is defined as the 'accomplishment or carrying out of a command' (OED Online), there are two aspects that need to be addressed; firstly, the carrying out of an activity or task, and secondly, the outcome of the completed task. Implicit in this definition of performance is the existence of performers, instructors and beneficiaries of performance. Whilst these may be the same person or a multitude of different stakeholders, there is a variability of performance in which each performer is different in terms of ability and experience; the instructor has influence in terms of determining the objective of the task and identifying the performance level expected from the performer, and the experience of beneficiaries is determined by their needs and expectations. Performance is also influenced by the objective of the task and can be relative. Therefore, when performance is reviewed, it is necessary to appreciate the different aspects and factors that influence performance.

Neely (1998) defines performance measurement as 'the process of quantifying the efficiency and effectiveness of past actions'. This implies performance will not always need to be measured. For example, there may be no need to assess the performance of carrying out a task, if the task is straightforward and not repeated. Performance should only be measured when past action can influence current or future action.

Service profile

To measure performance it is necessary to understand the activities and resources required to achieve the completion of the task. Effective measurement is based on metrics that assess both the carrying out of the task (input metrics) as well as its completion (output metrics). Performance is measured to provide information to analyse trends in performance, improve efficiency and effectiveness of future actions, compare different performers, reduce uncertainty of future actions and predict future performance.

Neely (1998) outlines three functions of business performance measurement:

1. Compliance: to ensure its operations comply with non-negotiable obligations.
2. Checking: to ensure intended operations are being undertaken efficiently and effectively.
3. Challenge: to question whether the intended operations are the appropriate use of the resources available.

The first function helps to ensure that the operations of the organisation comply with all necessary mandatory and legal obligations. The second function indicates the need to check the comparative position of one organisation against another. For example, organisations may be subject to independent external measurement to assess their comparative performance. Regulators assess privatised companies and government assesses public authorities, including the performance of their schools and hospitals. These measures often provide a clear indication of the performance from a user's point of view. For the individual organisations such measurement could be an important incentive to challenge how their existing operations are undertaken. The third function of measurement is as a stimulus for improvement.

Clearly, companies need to measure any aspect of their performance that can influence current or future action. But as Eccles (1991) highlighted, organisations traditionally assessed their performance on a purely financial basis using indicators such as return on investment, profitability and turnover. However, according to Kaplan and Norton (1996), measuring a company's performance based on its financial measures alone is no longer sufficient and a broader approach to performance measurement led to the development of their management theory called the Balanced Scorecard. The Balanced Scorecard is a framework to translate an organisation's mission and strategy into tangible objectives and measures. The principle of a Balanced Scorecard is to construct a comprehensive set of measures which identify the critical areas of the business. Whilst the Balanced Scorecard may retain an emphasis towards financial objectives, it also incorporates 'non-financial' measures. The non-financial measures are fundamentally the performance drivers behind the financial objectives. An example of a Balanced Scorecard is provided in Figure 9.7.

Rethinking Construction encouraged the construction industry to adopt performance measurement. Egan's report stated that 'the industry must replace competitive tendering with long-term relationships based on clear measurement of performance and sustained improvement in quality and efficiency' (Egan, 1998, p.5). The report identified the scope for 'sustained' improvement in the construction industry by listing measures of performance. Table 9.1 shows Egan's seven key performance indicators (KPIs) and their potential scope for annual improvement.

The Department of Environment, Transport and the Regions (DETR) in conjunction with the Construction Best Practice Programme (CBPP) developed a Key Performance Indicator pack, based on the indicators above, which provided a method for benchmarking performance against the KPIs. Companies utilised the wallcharts and guidance supplied and submitted figures for collation into the annual statistics for the industry. The successor to the CBPP, Constructing Excellence continues to collate the annual statistics for those companies that subscribe to the service. The *UK Industry Performance Report 2014* (see Figure 9.1) has amassed a reasonable dataset that has enabled certain trends in performance to be assessed.

Whilst the work undertaken as a result of Egan's performance targets for improvement has raised the profile of performance management, the adoption of performance measurement has not been universal across the industry. Moreover, trend analysis generated from the KPI report does undeniably indicate improvement, however it is only based on information from companies that submit the data, and therefore has a bias towards companies that have measured performance.

Table 9.1 The scope for sustained improvement.

Indicator	Improvement per year
Capital cost	Reduce by 10%
Construction time	Reduce by 10%
Predictability	Increase by 20%
Defects	Reduce by 10%
Accidents	Reduce by 20%
Productivity	Increase by 10%
Turnover and profits	Increase by 10%

Source: Egan, Sir J, (1998) *Rethinking Construction The Report of the Construction Task Force*

Figure 9.0 UK Industry Performance Report (2014)

In terms of Neely's three functions of measurement; to comply, check and challenge, Egan's performance targets are more focused on challenging outcomes in order to improve, with the exception of indicator 7 (turnover and profits), which is company specific. Whilst it is useful to have information on time and cost, (the first and second KPIs in Table 9.1), there is no indication of the source of any failure to meet the target, or how to remedy the situation in the future. This has long been a criticism of cost and time as measures of performance. Meyer (1994) for example, states, 'the fact that the project was late and over budget does not tell anyone what went wrong or what to do next'.

To achieve improvement in performance, it is necessary to have continuity. However, construction projects are often one-off and by their very nature have a temporary and finite life. Neither characteristic is conducive to achieving continuous improvement in performance. It is also important to recognise that the objectives of a construction project are fundamentally the objectives of the client, not the contractor or any of the consultants.

To realise the construction project, the client invariably needs to procure the services of a multitude of firms from within the construction industry to form a temporary project coalition. The firms in the coalition tend to have survival ambitions that are of a longer-term nature than the construction project. Rather than focusing on indicators for projects, it could be argued that the focus should be on developing performance indicators for the companies that form the project coalition. Well-managed projects are more likely to occur if the project coalition is a collection of well-managed companies. Hence measuring company performance should provide a more effective basis for achieving sustained improvement in construction.

The *Government Construction Strategy* (Cabinet Office, 2011) sets out the performance expectations of a significant client to the industry, namely the public sector. The clear objective of the strategy is to reduce construction costs by improving the performance of the industry. The strategy contains 13 themes (Figure 9.2) to improve performance, one theme being BIM. There are two key aspects of the strategy that are of particular significance for performance management. Firstly, there is recognition that the public sector has not managed its data requirements, including asset data effectively and this, in part, has affected its ability to acquire what it needs at the right price. The strategy urges clients to understand what projects should cost by maintaining appropriate benchmarking data. Secondly, the strategy emphasises the need for greater collaboration and hence the need for improved supplier relationship management, supported by effective performance assessment. One of the 13 themes in the strategy is the development of improved supplier relationship management, which includes the need for ongoing supplier performance assessment.

The implementation of the *Government Construction Strategy* has focused on four principal elements. These are new procurement models,[1] BIM, soft landings and a more intelligent (informed) client[2] (Cabinet Office, 2012, 2013). All four elements highlight the desire for effective performance management. The report produced by the Procurement/Lean Client Task Group (Cabinet Office, 2012) focused on new procurement models and the Informed Client (see Chapter 5). According to the report, the underlying principles of an Informed Client included recognition of the need to develop a collaborative culture between client and supply chain with a commitment to continuous improvement. Subsequent case

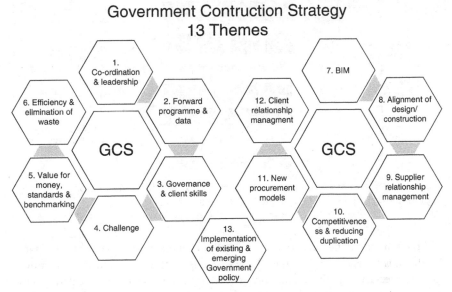

Figure 9.2 Thirteen themes of the *Government Construction Strategy.*
Hughes (2013)

studies of trial projects have demonstrated the characteristics of an Informed Client employing the new procurement models. For example, in the case study of Supply Chain Management Group (SCMG)[3] (Cabinet Office, 2014), the client adopted a two-stage open book approach, which encourages a more collaborative working approach supported by effective performance management. Alongside improvements in cost and time performance, the numerous qualitative benefits included improved specification and more efficient procurement, accelerated briefing reducing risks, early engagement of supply chain enabling innovation, sharing of best practice as well as facilitating the recruitment of local skills and taking on apprenticeships. Most significantly, the improved relationship between client and supply chain enabled higher levels of productivity and performance to be achieved.

BS 8534:2011 Construction procurement policies, strategies and procedures defines performance management as

> a systematic approach to help manage and lead the delivery of projects, linking business goals developed in the planning and development stages to overall project outcomes in the implementation, operation and decommissioning stages. The approach ensures that project success is clearly defined, is measurable and is ultimately achieved (BSI, 2011).

A performance management system is promoted which communicates strategy and receives feedback on achievement; the system is represented in Figure 9.3.

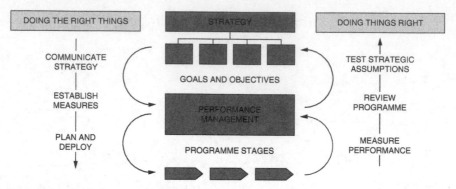

Figure 9.3 Performance management system BS 8534:2011: construction procurement, policies, strategies, and procedures.
BSI (2011)

Whilst it could be argued that the performance targets set out in Egan's *Rethinking Construction* might have been unrealistic, the report did give the industry the necessary impetus to focus on performance improvement. The demand for greater measurement of performance, which is in part due to the enhanced capability of information systems, has seen an increasingly sophisticated approach to project controls,[4] as demonstrated in the case studies below. Informed clients display clear leadership and recognise the need for effective measurement of both inputs and outputs to drive overall improvement. When the emphasis of performance management is focused on organisations rather than projects, improvements are more likely to be sustained.

Case study details

Case studies are presented on two organisations that have adopted an integral approach to managing the performance of construction projects. Firstly the Ministry of Justice (MoJ) will demonstrate how the adoption of, initially simple measurement of cost and time, enabled greater transparency over performance that resulted in the organisation being at the forefront of BIM implementation. Secondly, Crossrail is shown as a client that has employed a performance management framework predicated on effective information management principles to assure trusted inputs and produce outputs to ensure compliance, check progress and drive efficiency and best practice across the supply chain. The case studies provide examples of services developed around performance management, and for which the QS is ideally skilled to deliver.

Ministry of Justice

The Ministry of Justice (MoJ) spends approximately £300m per annum on maintenance and new-build projects, making it the Government's largest centrally mandated estate. In 2009 the MoJ embarked on the initiative Transforming

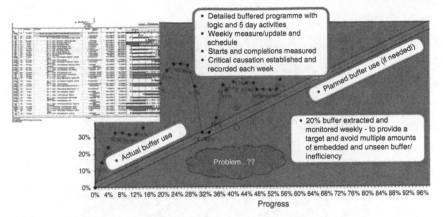

Figure 9.4 MoJ Cookham Wood performance measurement.
Buildsoffsite (2013)

Justice 'aimed at achieving a better justice system at less cost to the public (Gash & McCrae, 2010). Stocks (2011) defined the vision and strategy of this initiative, with its focus on improvement, and explained how this influenced the Estates Directorate to develop a systems approach to its project management; the system generated a clear set of benchmarking data, i.e. comparing project delivery against budget and time, providing transparency over performance as well as being a tool to support the creation of a lean culture.

The evidence presented by Stocks (2012), at a subsequent event, illustrated that the MoJ was able to demonstrate significant improvements in its performance against National Construction Industry Key Performance Indicators. As a consequence of the improvement initiative, the MoJ sought a technology solution that would provide the organisation with better information just as BIM was being introduced. Hence, the MoJ was an early adopter of BIM and well placed to provide Cookham Wood – a male juveniles' prison and Young Offenders Institution in the village of Borstal in Kent – as a trial project forming part of the Government Construction Strategy Implementation Plan.

The case study of Cookham Wood highlights the use of performance measurement and management (Buildoffsite, 2013). Figure 9.4 illustrates one of the performance charts used to track activity from the master programme with weekly recording of actual versus planned completion. When activities were not completed as planned, data was collated to identify the reasons for non-completion and allow further analysis of trends.

Crossrail

Crossrail, a £14.9bn programme creating a new rail infrastructure for London and the South-east, required the passing of an Act of Parliament, and as a result, the monitoring of project progress came under close scrutiny. Information

Figure 9.5 Categories for assessing Tier 1 suppliers.
Provided courtesy of Crossrail.

management proved a key component of the Crossrail approach.[5] Whilst Crossrail produced a major report on a monthly basis for the many different stakeholders, the project leaders recognised the importance of collating reliable data to ensure the report was trusted. Because many sources provided Crossrail with data, an assurance process was established involving a structured hierarchy to check data quality. The assurance process ensured data was entered correctly in the first place and also checked the data sources themselves to confirm confidence in the information.

Moore (2013) highlighted that Crossrail has invested in, 'a robust performance framework that has driven efficiency and best practice across the supply chain'. Figure 9.5 identifies the headline categories that were used to evaluate the performance of Crossrail's Tier 1 suppliers. As a result of the assessments, the relative performance of the supply chain improved by 44 per cent and enabled the National Audit Office to conclude 'that Crossrail is well on target to offer value for money' (Moore, 2013).

Salih & Vale (2014) explained how Crossrail had identified that the point of failure in performance is generally a consequence of human error, thus the need for improved and consistent data. This was achieved by embedding appropriate information systems into supply chain partners in order to ensure commonality of data processes. Crossrail therefore invested in the Information Academy to assist in developing understanding and capability in the organisations with

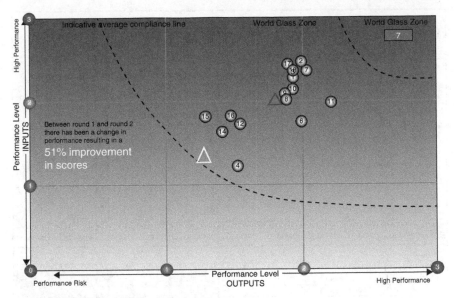

Figure 9.6 Crossrail contractor performance.

whom it contracted. The resulting project reports provided assurance, a performance check and a comparison of Tier 1 contractors to generate a league table of performers. This league table created the performance challenge and incentivised contractors as no contractor wanted to be at the bottom of the league (Construction News, 2013). Figure 9.6 illustrates performance between two rounds of assessment. It demonstrates that measurement can motivate significant improved performance.

Figure 9.7 provides a further example of how contractor performance is monitored and reported on Crossrail.

As can be seen from these two case studies, performance management can take many forms, and it will almost certainly be determined by the nature and requirements of the client's business. It will involve both the measurement and analysis of data – core skills of the QS, along with knowledge of databases, auditing and benchmarking to assess performance against agreed targets. Whilst performance management may be commissioned as a separate service, it is probably more likely to form part a wider Project Management brief – or perhaps even forming part of an Information Management role. Moreover, with increasing emphasis on adopting more collaborative forms of working, there is a shift from a focus on cost metrics to a broader set of metrics that encompass the performance of a project. This is likely to become even more pertinent with increasing adoption of Social Value by the public sector (Cabinet Office, 2015).

So what might this role involve? Table 9.2 below offers an outline to assist in determining the scope and nature of performance measurement and management services to be provided on a project:

Technical Information Dashboard - Period 01

Contract Performance Summary (Crossrail)		Modelling					Documention & Data					GIS			Asset Info			
		Contractor Average	CAD Support	CAD Management	CAD Data Publishing Compliance	Use of 3D|4D for Design & Construction	Right First Time	% of Deliverables completed	CRLDMT Support	CRL Health Check	Contractor Health Check	Accessibility	Degree of use	Volume of use	Contractor Engagement	ADCS Submission Response Time	Population of Asset Tag names	Data Quality
Parameter Average		3.3	3.5	3.4	4.2	2.5	2.9	4.1	4.6	3.6	3.9	2.5	2.4	3.7	4.1	3.9	4.3
Ref. Contractor / Contract		●	●	●	O	●	O	O	X	X	●	O	●	●	●	●	●

Owner — Cross rail Ownership X; Joint Ownership O; Contractor Ownership ●

Scoring — 1-2 Poor; 3: Concern; 4-5: Good; World Class; N/A/ TBC

CIVILS

Ref.	Contractor	Contract	Contractor Average	CAD Support	CAD Management	CAD Data Publishing Compliance	Use of 3D|4D	Right First Time	% of Deliverables completed	CRLDMT Support	CRL Health Check	Contractor Health Check	Accessibility	Degree of use	Volume of use	Contractor Engagement	ADCS Submission Response Time	Population of Asset Tag names	Data Quality
1	Contract 1	Civils Contract 1	3.5					2	3	5	5	5	3	1	1	4	4	4	5
2	Contract 2	Civils Contract 2	3.6	3				4	4	4	5	4	4	3		4	3	4	5
3	Contract 3	Civils Contract 3	4.2	2				4	4	5	5	4	4	4	5	4	5	4	5
4	Contract 4	Civils Contract 4	4.1	5				3	4	2	5	4	4	4	4	5	4	5
5	Contract 5	Civils Contract 5	3.9	4				4	4	5	5	5	4	2	1	4	5	4	4
6	Contract 6	Civils Contract 6	2.7										4	2	2				
7	Contract 7	Civils Contract 7	3.3	2	4	4		2	3	5	5	4	4	3	1	3			
8	Contract 8	Civils Contract 8	3.8	3	4	5	4	3	4	5	5	4	4	3	4	4	2		
9	Contract 9	Civils Contract 9	3.7	2	3		5	3	5	5	4	4	3	3	3	4	5		
10	Contract 10	Civils Contract 10	4.5					4	5	4	5								
11	Contract 11	Civils Contract 11	2.8					4	5	1	1								
12	Contract 12	Civils Contract 12	3.8					4	4	5	5	4	4	2	2				
13	Contract 13	Civils Contract 13	2.9	3	4	3		1	2	2	5	4	3	1					

Figure 9.7 Benchmarking data applications and contract performance.

Table 9.2 Indicative scope of performance measurement and management services.

Scope	Service delivery
Strategic	Identify/agree client objectives. Identify/agree what and how this should be measured. A Balanced Scorecard is likely to form the basis of this, either existing/standard or developed specifically for the project/client.
Operational / data analysis	Measurement of all aspects of the Balanced Scorecard. Data analysis, understanding when to act and who is responsible. The challenge is to understand what data is required that drives the required behaviour to deliver performance improvement.
Auditing / benchmarking	May form part of a quality management system. Monitoring and assessing against agreed benchmarks.

How BIM could be used

BIM can facilitate the further development of performance management. Since it was advocated in *Rethinking Construction* in 1998, performance management has evolved through adoption on large programme management projects, such as the London Olympics 2012 and Crossrail, and through other initiatives such as lean construction (Stocks, 2012 and Terry and Smith, 2011). This evolution has seen an increasing importance in the role of performance measurement, with the recognition of the need for collating appropriate data and an increasing emphasis on effective management of information. However, with improved data collation and enhanced utilisation of the available technology, BIM should facilitate further advances in the evolution of performance management, for example, the use of digital tracking and monitoring techniques.

The opportunity BIM presents is significant, and it has the potential to lead to radical improvements as envisaged by Latham and Egan. This opportunity to change the way construction projects are managed depends largely on the effective use of data to create more efficient ways of working and to develop more appropriate solutions for clients. This is recognised in the publication of PAS 1192-2:2013 specification for information management for the capital/delivery phase of construction projects using building information modelling (BSI 2013), setting out the industry standard for managing information. It will be necessary to set out in the Employers Information Requirements (EIRs) the information required to manage the performance of the project and the completed asset, and to ensure that these requirements are subsequently reflected in the post-contract BEP (refer to Chapter 5 for more on the EIR and BEP as part of the procurement process).

Earlier, this chapter referenced three functions of business performance measurement; checking progress, ensuring compliance and measuring performance for improvement. As evident from the case studies featured in this chapter, BIM can be used to fulfil these functions. Firstly with respect to compliance, BIM creates a digital model of a built asset that can be assessed for compliance with the multitude of building and other regulations that affect construction. The data embedded in a digital model may be used by planning officers, building control and other authorities to verify that a model complies with planning, building regulations and even health and safety (Mordue, 2012). Software is already available for checking fire escapes, whereby a building control officer is able to verify that all fire escape routes comply with building regulations, quickly and easily, and it is reasonable to assume that any set of rules could be validated for compliance. This removes significant manual processes, saving time and cost.

Secondly, BIM also offers new opportunities to monitor progress of both design and construction phases of a project. It is well understood that the design phase, whilst being a small proportion of the total cost of ownership, has a significant influence on the final cost of the product (Evans et al., 1998 and Saxon, 2005). Nevertheless, the quality of the design phase has remained difficult to assess (Coates et al., 2010). Essentially, one measure of the quality of the design process relates the time involved in the design phase to the delivery of appropriate outputs. BIM permits greater transparency during the progress of design work, representing the completion of design information in relation to the design programme. Whilst this in itself is not a measure of the quality of the output, the subsequent use of tools, such as those employed to aid clash avoidance will provide a degree of measurement of design quality. Light (2011) describes the concept of the MacLeamy Curve, whereby the effort of the design phase is shifted to encourage greater integration of project design and delivery. Coates et al. (2010) assessed the effectiveness of adopting BIM in a design practice, acknowledging the importance of developing metrics to measure the quality of projects, information and design activity. They considered these metrics assisted in justifying the return on investment of BIM and argued that adopting BIM allowed design practices to add value to their designs.

There are also significant opportunities to improve the measurement of progress during the construction phase. Eastman (2011) identifies the development of potential digital tracking and monitoring techniques, which are moving from theory to application with the benefit of having precise and timely information to facilitate improved decision making. These technologies include radio-frequency ID (RFID) tags adopted by Laing O'Rourke (n.d.) during the delivery of the Leadenhall Building in the City of London. Data tags attached to building components can be tracked through manufacture, supply and installation, providing up-to-date information on progress to plan possible preventative action should delays arise. Moreover the same data may be used to facilitate commercial activities such as valuations and payments. Laing O'Rourke state that 'when integrated with BIM, RFID can be used to render a data-rich replica of the project in real time'. They go on to say 'this technology will be used to enhance project controls' and 'develop robust key performance indicators'. This creates the possibility, together with laser scanning, of project reporting changing from being mainly text based to becoming a dynamic 3D visual representation of progress.

Recent high-profile projects such as the Olympic and Paralympic Games in London 2012 demonstrated the ability of the construction industry to perform. London 2012 has been a source of pride for politicians, commentators, contractors and the construction industry in general, given the challenges that were faced in completing the Commonwealth Games in India in 2010 and the plight of the Brazilian construction industry in building the venues for the FIFA World Cup in 2014. Mead and Gruneberg (2013) emphasise the importance of performance measurement, and the use of a Balanced Scorecard in supporting the successful completion of London 2012. The authors describe the development of a 'framework for measuring and management performance depends on defined criteria for success' (p. 50) and show how the Balanced Scorecard aligns the criteria for success with the client's mission statement through the programme objectives, underpinned by a standard set of critical success factors (Figure 9.7).

Tools capability

Some suppliers of collaborative document, data and project management systems, provide functionality to support performance measurement. These tools utilise the data within a model to generate project information that facilitates

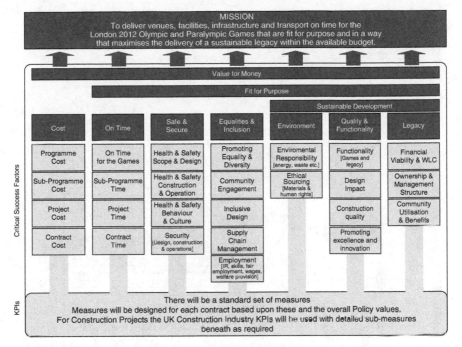

Figure 9.8 Olympic Delivery Authority, Balanced Scorecard.
Mead and Gruneberg (2013)

decision making, as they are capable of monitoring and tracking performance through the various stages of a project, with information generated either in real time or for regular project control reports. The information can be generated in a variety of different formats, such as an online dashboard (Figure 9.9). Moreover, in-depth analysis of data enables performance metrics to be compared and contrasted at various levels of a project, highlighting key trends as well as identification of non-compliance or failure. Another key attribute is with greater transparency of the data, the management of risk and change is enhanced by understanding of the effect of change on all aspects of the project, supporting a 'right first time' approach.

Issues/benefits

The principal issues and benefits of managing performance with BIM tools are aligned to the industry's capability to measure performance. Basic project reports, such as monthly cost and progress reports are forms of performance measurement, and take on the performance role of 'checking' during pre-contract phases or during the construction phase. As a standard component of the project management process, these reports are used to assess how a project is progressing on a regular basis. As with Egan's performance targets, the emphasis of project reports is to focus on the monitoring of the current status of the standard indicators of time,

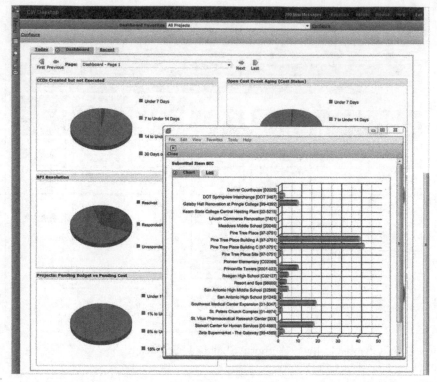

Figure 9.9 Dashboard illustration, Constructware.
Autodesk (2014b)

cost, quality and safety. However, project controls are evolving due to the increasing capability of information technology, to deliver the demand for greater information; document management and collaborative technologies are now common practice on many projects. This digital evolution of project controls has also forced project leadership to develop capabilities to understand what information is needed to manage projects, whilst improving the technical capability of the users and the technology itself. The challenge for the industry, and specifically quantity surveyors, is to ensure the necessary skills required to manage data efficiently and appropriately are present. It could be argued that there are already plenty of capable individuals within the industry with the necessary data management skills, but who are seen as slightly geeky and not particularly valued. Thus it may not necessarily be a case of skill shortage but more the case of perhaps identifying and utilising these undervalued skills to achieve competitive advantage.

The development of information technology (IT) systems has increased the digital information available which in turn enhances the ability to effectively manage performance in terms of compliance, checking and challenging. As outlined in both case studies from the Ministry of Justice and Crossrail, the IT systems support the performance framework, and this framework provides transparency over performance that can drive significant improvements both within projects and organisations.

The specific issue of measuring performance within BIM relates more specifically to the issue of interoperability. This means ensuring systems are capable of efficiently exchanging data as is discussed in Chapter 6. As Wilkinson (2007) highlighted, quoting the president of McGraw-Hill Construction, Norbert Young: 'The lack of seamless flow of information – or interoperability – is one of the primary factors holding the entire industry back from quantum leaps forward'. Interoperability is still an issue and it is imperative for effective performance management, that the performance products available work seamlessly, and not just within each software supplier's own suite of products. The potential benefits of interoperable systems could support significant enhancement in value for all stakeholders, with improved efficiency and effectiveness of working practices. This represents both an opportunity and threat for the QS as BIM tools could enhance their service offering, but could just as easily be utilised by other professionals.

When configuring a project, defining the performance management framework has to date not been seen as a priority. However, implementing BIM and utilising PAS 1192-2:2013 (Specification for Information Management using BIM) should ensure the information required to manage the performance of the project and the completed asset is established at the outset, forming part of the EIRs and subsequent BEP. A wider issue that is likely to reinforce the need for both greater performance measurement and the wider adoption of BIM is the focus on developing client capability and procurement as outlined in *Construction 2025* (HM Government, 2013) and the IUK Project Initiation Routemap (HM Treasury, 2014). It is recognised that traditional procurement practice focuses on price and output, rather than cost and outcome. Improving client awareness and capability will influence more collaborative working arrangements, which in turn will demand improvement in performance that can be provided by tools supporting a BIM approach. Further opportunity will arise from the switch from the Built Environment from the suppliers of assets to the suppliers of facilities (Saxon, 2005). This shift towards Built Environment as a Service (BEaaS) will place the emphasis on organisations to adopt effective performance management for the life cycle of the asset.

Summary/commentary

Whilst the primary output of BIM might well be considered to be the creation of a richer, deeper and re-usable pool of data, the primary outcome is almost certainly all about improving performance – whether that be in terms of the delivery process, the people or performance of the asset. The UK Government's targets embodied in both its 2011 construction strategy, and more recently, within its *Construction 2025* report (HM Government, 2013), are all about performance improvement, and this is likely to encourage more demand for value-added services related to the measurement and management of performance.

Performance management encompasses the full remit of a project and is an integral part of managing an organisation. Given that performance management involves the collation, monitoring and reporting of information, it seems that BIM will play an increasingly significant role in the continuing evolution of construction performance management. Moreover, the drivers for change within the industry, with improved client capability and a supply side establishing value-added service offerings, will demand more dynamic performance information. As Carder (2012) indicates, performance management is a key element of the Informed Client and should be carried out by the client or the client representative and not left to the supply side. The QS has the core skills to develop the service offering that encompasses the strategic and operational aspects of performance management.

There is tremendous scope for improvement in the use of performance management within construction, both on projects and within individual organisations. From a project perspective, managing performance is about ensuring compliance, greater control and improved real-time assessment of progress; whereas for organisations it's about improvement. Maher (2015) highlights incorporating simple measures into BIM to drive collaborative behaviours and a successful project: 'Success was linked to key measures that were revealed to all and these were kept simple, such as minimising the number of clashes in the virtual world, resulting in little rework on site.'

However, the starting point is measurement, and hence, with the increasing adoption of BIM, the QS has much to offer and gain from the development of performance management. Measuring performance will not, in itself, improve performance (Deming, 1986), so acknowledging that both performance management and BIM are simply methods of working and tools to facilitate the creation of the built environment, improvement in the built environment will come from better understanding and effective utilisation of these tools and the methods of working they bring with them. This will demand a better understanding of the information needed to manage projects, as well as improved technical capability of the users and this will challenge the traditional skills set of the QS.

Notes

1   Refer to Chapter 5 for more on the New Procurement Models.
2   Intelligent Client is the term used by the UK Government and adopted throughout this chapter, although the authors believe a more appropriate term to be Informed Client.
3   SCMG – Supply Chain Management Group is a multi-client, multi-contractor team engaged on housing refurbishment that has worked with a wide range of SME subcontractors and suppliers under a standardised system of costing and long-term engagement that has created major savings and significant qualitative benefits (HM Government).
4   See also Chapter 8 for more examples of how BIM can enhance project controls as part of Contract Administration.
5   Refer to Chapter 6 for further details on the Crossrail programme and the role of Information Management.

References

Autodesk (2014a) 'Buzzsaw: document and data management software as a service', Online. Available HTTP: <http://www.autodesk.co.uk/products/buzzsaw/overview>, accessed 22nd July 2014.

Autodesk (2014b) 'Constructware', Online. Available HTTP: <http://www.autodesk.com/products/constructware/overview>, accessed 22nd July 2014.

Barker, D. (2011) 'BIM – measurement and costing' NBS, Online. Available HTTP: <http://www.thenbs.com/topics/bim/articles/bimMeasurementAndCosting.asp>, accessed 7th March 2014.

Bentley (2014) ProjectWise Construction Work Package Server, Online. Available HTTP: <http://www.bentley.com/en-GB/Products/ProjectWise+Construction+Work+Package+Server/>, accessed 22nd July 2014.

BSI (2011) *BS 8534:2011 Construction procurement policies, strategies and procedures – code of practice*, British Standards Institute, London.

BSI (2013) *PAS1192-2 Specification for information management for the capital/delivery phase of construction projects using building information modelling*, British Standards Institute, London.

BuildOffSite (2013) 'HMYOI Cookham Wood: houseblock and education building', Build Off Site, Online. Available HTTP: <http://www.buildoffsite.com/pdf/publications/BoS_CookhamWood.pdf>, accessed 23rd July 2014.

Cabinet Office (2011) 'Government construction strategy', HMSO London, Online. Available HTTP: <http://www.cabinetoffice.gov.uk/resource-library/government-construction-strategy>.

Cabinet Office (2012) 'Final report to Government by the procurement: Lean Client Task Group', HMSO London, Online. Available at <https://www.gov.uk/government/uploads/system/uploads/attachment_data/file/61157/Procurement-and-Lean-Client-Group-Final-Report-v2.pdf>.

Cabinet Office (2013) 'Government Soft Landings section 1 introduction', London, Online. Available at <http://www.bimtaskgroup.org/wp-content/uploads/2013/05/Government-Soft-Landings-Section-1-Introduction.pdf>.

Cabinet Office (2014) 'Procurement trial projects case study report - Hackney/Haringey SCMG social housing refurbishment', London, Online. Available at <https://www.gov.uk/government/uploads/system/uploads/attachment_data/file/325951/SCMG_Trial_Projects_Case_Study__CE_format__130614.pdf>.

Cabinet Office (2015) 'Social Value Act review', London, Online. Available at <https://www.gov.uk/government/uploads/system/uploads/attachment_data/file/403748/Social_Value_Act_review_report_150212.pdf>.

Carder, P. (2012) 'Lead function: "Intelligent Client", or, "Strategic Sourcing"?', The Occupiers' Journal, Online. Available at <http://occupiersjournal.com/lead-function-intelligent-client-or-strategic-sourcing/>, last accessed 3rd April 2015.

Coates, P., Arayici, Y., Koskela, L., Kagioglou, M., Usher, C. and O'Reilly, K. (2010). The key performance indicators of the BIM implementation process. In *Computing in Civil and Building Engineering, Proceedings of the International Conference*, W. Tizani (Editor), 30 June–2 July, Nottingham, UK, Nottingham University Press, Paper 79, p. 157

Constructing Excellence, Glenigan, CITB and UK BIS (2014) UK Industry Performance Report 2014, Glenigan

Construction News (2013) 'Crossrail compares contractor performance to drive world-class standards', Online. Available HTTP: <http://www.cnplus.co.uk/home/news-analysis/crossrail-compares-contractor-performance-to-drive-world-class-standards/8656604.article#.U5md7bElknY>, accessed 12th March, 2014.

Dave, B., Koskela, L., Kiviniemi, A., Owen, R. and Tzortzopoulos, P. (2013) *Implementing Lean in Construction: Lean Construction and BIM*, CIRIA, London.

Deming, W.E. (1986) *Out of the Crisis* 1st MIT Press ed., 2000. edn. Cambridge, Mass.; London

Eastman, C.M. (2011) BIM handbook : a guide to building information modeling for owners, managers, designers, engineers and contractors, 2nd ed. Wiley Hoboken, NJ

Eccles, R.G. (1991) 'The Performance Measurement Manifesto', *Harvard Business Review*, January–February 1991, pp.131–137.

Egan, J. (1998) *Rethinking Construction: report from the Construction Task Force*, Department of the Environment, Transport and Regions, London.

Evans, R., Haryott, R., Haste, N. and Jones, A. (1998) *The Long Term Costs of Owning and Using Buildings*, Royal Academy of Engineering, London.

Gash, T. and McCrae, J. (2010) 'Transformation in the Ministry of Justice', Institute for Government, Online. Available at <http://www.instituteforgovernment.org.uk/sites/default/files/publications/Transformation%20in%20the%20Ministry%20of%20Justice%201st%20Interim%20Report.pdf>, accessed 23rd February 2014.

Green, S.D. (2011) *Making Sense of Construction Improvement*, Wiley-Blackwell, Chichester.

Harris, EC., LLP (2013) *Supply chain analysis into the construction industry*, A report for the Construction Industrial Strategy, research paper No. 145, Department for Business, Innovation and Skills, London, Online. Available at <https://www.gov.uk/government/uploads/system/uploads/attachment_data/file/252026/bis-13-1168-supply-chain-analysis-into-the-construction-industry-report-for-the-construction-industrial-strategy.pdf>.

HM Government (2013) 'Construction 2025 Industrial Strategy: Government and Industry in Partnership BIS/13/955', London, Online. Available at <https://www.gov.uk/government/publications/construction-2025-strategy>.

HM Treasury (2014) 'Improving Infrastructure Delivery: Project Initiation Routemap Handbook', Online. Available at <https://www.gov.uk/government/uploads/system/uploads/attachment_data/file/361173/0208_Routemap_Handbook_30_Sept.pdf>, last accessed 3rd January 2015.

Hughes, G. (2013) *Government Construction Strategy*, unpublished BSc presentation, University of Westminster, London.

Kaplan, R.S. and Norton, D.P. (1996), *The Balanced Scorecard Translating Strategy into Action*, Harvard Business School Press, Boston.

Laing O'Rourke (nd) 'The Leadenhall Building. London. UK', Online. Available HTTP: <http://www.laingorourke.com/our-work/all-projects/the-leadenhall-building.aspx>, accessed 7th March 2014.

Latham, Sir M. (1994) *Constructing the Team*, HMSO, London.

Lean Construction Institute (n.d.) 'The Last Planner (R)', Lean Construction Institute, Arlington, Online. Available at <http://www.leanconstruction.org/training/the-last-planner/>, accessed 28th January 2014.

Light, D. (2011) 'BIM Implementation – HOK building SMART', NBS, Online. Available at <http://www.thenbs.com/topics/bim/articles/BIM-Implementation_HOK-buildingSMART.asp>, accessed 7th March 2014.

Maher, A. (2015) 'BIM drives valuable collaboration', Arup Thoughts Blog, Online. Available HTTP: <http://thoughts.arup.com/post/details/407/bim-drives-valuable-collaboration>, accessed 3 April 2015.

McCabe, S. (2001) *Benchmarking in Construction*, Blackwell Science, Oxford.

Mead, J. and Gruneburg, S. (2013) *Programme Procurement in Construction, Learning from London 2012*, Wiley-Blackwell, Chichester.

Meyer, C. (1994) 'How the right measures help teams excel', *Harvard Business Review*, May–June 1994, p. 95.

Moore, P. (2013) 'Half way there, but no half measures – lessons learned from Crossrail's success', Turner and Townsend, UK, Online. Available HTTP: <http://www.turnerandtownsend.com/lessons-crossrail/_21826.html>, accessed 5 September 2014.

Mordue, S. (2012) 'CDM and BIM – virtually sorted', NBS, Online. Available HTTP: <http://www.thenbs.com/topics/bim/articles/CDM-BIM-Virtual-sorted.asp>, accessed 7th March 2014.

Muse, A. (2013) 'BIM will be a Trojan horse for revolution and change in the construction industry', *Sustain Magazine*, Online. Available HTTP: <http://sustainmagazine.com/bim-will-be-a-trojan-horse-for-revolution-and-change-in-the-construction-industry/>, accessed 9 February 2014.

Neely, A. (1998) *Measuring Business Performance*, The Economist Books, London.

Neely, A. (2002) *Business Performance Measurement: Theory and Practice*, Cambridge University Press, Cambridge.

OED Online. March 2014. Oxford University Press, Online. Available HTTP: <http://www.oed.com/view/Entry/140783?redirectedFrom=performance>, accessed 7 March, 2014.

Salih, H. and Vale, P. (2014) 'Interview with Crossrail on Performance Management', London 31 January 2014 [Hasan Salih is Right First Time & Reporting Manager and Peter Vale is IM Best Practice Manager].

Saxon, R. (2005) *Be Valuable*, Constructing Excellence, London.

Stocks, T. (2011 November 22nd) *Construction Strategy BIM Implementation, A Client View* [PowerPoint slides] presented at CIRIA event 'How can BIM make you more productive and profitable?' at Arup Offices, London.

Stocks, T. (2012 March 1st) *Lean in Construction, A Construction Client View* [PowerPoint slides] presented at CIRIA event 'Transforming the construction sector: what does Lean Thinking have to offer?' at CIRIA Offices, London.

Terry, A. and Smith, S. (2011) *Build Lean: Transforming Construction using Lean Thinking*, CIRIA, London.

Trimble, (2014) 'Work management', Online. Available HTTP: <http://www.trimble.com/fsm/work_management.aspx?tab=Performance_Insight>, accessed 22 July 2014.

Wilkinson, P. (2007) 'Software incompatibility bar to interoperability', Online. Available HTTP: <http://extranetevolution.com/2007/10/software-incomp-2/>, accessed 3 January 2015.

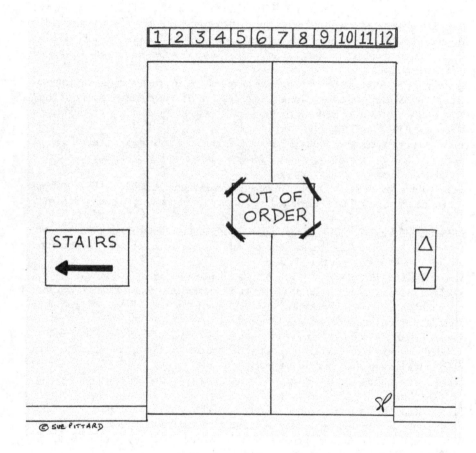

© SUE PITTARD

10 Facilities management

Graeme White and Phil Boyne

Introduction

The Government's work to date has concentrated on the construction and remediation of assets. In undertaking this work, and from emerging industry feedback, the value of BIM data to bring major efficiencies to the maintenance, operation and energy management of constructed assets is becoming clear.

Industry and UK Government believe that the cost savings and other benefits stemming from the use of BIM data will dwarf that generated in the construction of the asset. Over a 30-year life, the cost of operating the asset could be as much as four times the cost of designing and constructing the building (i.e. 80 per cent of the whole life cost).

Given this potential for significant savings here, it is important that those involved with the creation and use of information leverage the benefits that structured data can bring to the functionality, operational efficiency and effectiveness of assets, and their impact on the built environment.

The handover of assets from construction to operation is a critical activity, to ensure that the 'as designed' performance is achieved as soon as possible and that the on-going operation continues to conform to the planned 'as designed' parameters.

Facilities Management (FM) is one area identified in the UK *Government Construction Strategy* (HM Government, 2011) and it is being supported through the implementation of Government Soft Landings (GSL), which is now embedded as part of the project information delivery cycle for all central government projects to improve the transition from construction to operation of assets.

Directing the focus of the supply chain to deliver the required handover outcomes (GSL), together with long-term performance of assets is vitally important, and BIM data offers the information infrastructure or 'life blood' for such an approach. There is a strong synergy between data derived during delivery and the effectiveness and efficiency of the handover process, along with the operational evaluation of performance.

Author biographies

Graeme White

Graeme is a Chartered Engineer, a Fellow of The Chartered Institution of Building Services Engineers (FCIBSE) and a Fellow of The Institution of Mechanical Engineers (FIMechE).

He is an M&E Associate with Rider Levett Bucknall UK Limited and heads up their building services engineering division.

He has extensive experience of the design, specification, procurement, implementation and management of facilities management installation projects and operational contracts in both public and private sectors.

Phil Boyne

Phil is a Chartered Quantity Surveyor, with over 30 years of experience in the Construction and Facilities Management (FM) sector. He is a Fellow of the Royal Institution of Chartered Surveyors (FRICS) and a certified member of the British Institute of Facilities Management (CBIFM).

He has specialised in construction cost planning and estimating, working for cost consultants McGill & Partners, Gleeds and Faithful & Gould before joining Lend Lease Facilities Management in 2002, which was acquired by Cofely GDF Suez in July 2014.

During his appointment with Lend Lease FM, Phil has applied his construction cost planning and structured quantity surveying approach used for more traditional estimation and cost analysis, to the FM sector.

Company information

Rider Levett Bucknall (RLB)

RLB is an independent, global property and construction practice with over 3,500 employees in more than 120 offices across Asia, Oceania, Europe, Middle East, Africa and the Americas. Services provided include Cost Management, Project Management and Advisory Services. In addition to specialised advisory services, the firm maintains a cost-conscious and environmentally sustainable approach to business.

Cofely

Cofely, a GDF Suez company, is a European leader in energy, technical and facilities management services. The company develops innovative solutions that improve the efficiency of cities, buildings, industry and infrastructure.

Cofely in the United Kingdom and Republic of Ireland (UK & ROI) has a turnover of £1bn, and employs over 15,000 people. Cofely operates on 14,000 customer sites throughout the UK and ROI, totalling over 23.6m square metres of managed space.

Cofely is one of the five business lines of GDF Suez. GDF Suez Energy Services had 90,000 employees and revenues of €14.7bn in 2013.

Service profile

The British Institute of Facilities Management (BIFM) has adopted the definition of FM provided by CEN the European Committee for Standardisation and ratified by BSI British Standards: 'Facilities Management is the integration of processes within an organisation to maintain and develop the agreed services which support and improve the effectiveness of its primary activities'.

Effective FM, combining resources and activities, is vital to the success of any organisation. At a corporate level it contributes to the delivery of strategic and operational objectives. On a day-to-day level, effective FM provides a safe and efficient working environment which is essential to the performance of any business – whatever its size and scope.

FM is typically classified in terms of hard and soft services. Hard FM is the maintenance of the fabric, engineering services, landscaping and similar elements of an asset. The Royal Institution of Chartered Surveyors (RICS) indicates that effective maintenance planning and organisation of these services should cover the following:

- Planned preventative maintenance.
- Reactive maintenance (breakdown and repair).
- Emergency support.
- Predictive maintenance.
- Statutory maintenance.

Hard FM services would normally be required to meet an obligation to maintain an asset or facility. All of these activities relate effectively to the organisation's asset/property register and recognise a range of maintenance priorities, which are included within an annual programme of work. In this way the organisation's assets, both equipment and property, will be functional, safe and maintained over the life of the asset.

Soft FM covers services other than building and engineering which support the operation of the facility, typically cleaning, catering, portering, grounds maintenance, waste disposal and recycling, security, pest control, car parking, health and safety, cash collection, furniture and equipment, vending equipment and window cleaning.

One of the difficulties facing owners, tenants and facility managers is in knowing *what* information and supporting data is necessary to comply with a raft of legislative, financial, commercial, technical and management demands. The lack of available information can also have health and safety implications for owners and users.

In any FM operation, the effective delivery of service contracts will be a priority. Where multiple services are provided most large organisations (both public and private sector) will often have an in-house FM department in place to co-ordinate and manage the contracts. This will be particularly important where multiple services are provided by specialist service providers.

In all cases, there are likely to be service level agreements (SLAs) in place between the service providers, be they outsourced or in-house, and the client organisation. SLAs provide a set of criteria by which the performance of a contract may be assessed and managed to ensure compliance with client requirements.

Effective contract management processes ensure that all parties to the contract execute their obligations as efficiently and effectively as possible to satisfy the business and operational objectives, and particularly to provide value for money, while at the same time ensuring that the facility owner is at all times compliant in both statutory and contractual terms.

Traditional scope of service

The term of an FM contract is usually set, and in the case study example featured later in this chapter, it is set at 30 years, requiring the FM contractor to take a long-term view in terms of service provision, and also investment decisions with regard to service methodologies, equipment and approach. The common driver is to maintain the facility at a required standard over this term to ensure continued, 24/7 use by the client.

For new build, mobilisation takes place during the later stages of the construction phase to be ready for operation at the point of handover from the construction contractor.

Depending on the size or complexity of the facility, FM services may be organised and managed in a variety of ways ranging from simple paper-based systems and spreadsheets, through to integrated software packages employing handheld devices to provide real-time reporting.

During the mobilisation phase, typically three to six months, the focus will be on the necessary prerequisites for a successful commencement to the operational service, including the following:

- Gaining and compilation of building asset knowledge.
- Formation of a management-efficient service delivery model.
- Preparation of a stakeholder-focused service, to provide a speedy and effective response to maintenance issues.
- Pre-planning of delivery to meet Planned, Preventative Maintenance (PPM) requirements and Reactive Maintenance (RM) requirements to have been considered and planned for.
- Gaining of prior knowledge of asset maintenance requirements, operational skill requirements, hazards, access restrictions, mechanical and electrical system interfaces and resilience.

- Clarification of asset functions, work requirements, criticality and client service impact.
- Compiling and noting the PPM tasks required for all assets.
- Anticipating RM requirements and ensuring adequate asset knowledge is gained in order to be able to effectively deal with any such RM requirements.
- Efficient allocation of resource to PPM tasks and anticipated RM tasks.
- Setting out of performance monitoring requirements and the agreement of an effective system to enable such performance monitoring to take place efficiently, thereby allowing effective management of service levels during operation.
- Clear prioritisation of the effort required upon operational commencement to ensure the contracted FM service level is provided in accordance with stakeholder expectations and requirements.

Traditionally, information at project handover has often been piecemeal from disparate sources, such as 2D drawing files, PDFs, or in many cases boxes of paper drawings and documents. The end result is that the operations team have to spend unnecessary and significant time collating and validating the data for input into the FM system.

A significant part of the mobilisation phase is, therefore, to establish the necessary baseline for FM service commencement, including the compilation of a comprehensive asset database and register for the facility, drawing together construction details (fabric, mechanical and electrical services, fittings and equipment and external works), operational requirements, technical function, design, performance parameters, resilience and Information Communication Technology (ICT) capability. In a traditional environment this can be extremely labour-intensive and time consuming.

An asset database will provide the framework within which the FM contractor will store the key data required in order to ensure each asset is maintained as required under the contract.

Once data has been captured, it can be used to track and maintain life cycle information about the asset fabric (walls, floors, roof, etc.) as well as the equipment serving the building or facility (mechanical, electrical, plumbing, etc.) to plan and schedule a programme of maintenance activities that will improve asset performance, reduce repairs and reduce overall maintenance costs. In turn, this will allow facilities managers to plan maintenance activities proactively and appropriately allocate maintenance staff, as well as reducing corrective maintenance and emergency maintenance repairs. Using this information, facilities managers can also evaluate different maintenance approaches, analyse data to evaluate and make repair or replacement decisions and document the effectiveness of the maintenance programme.

Typically, FM services would be delivered to support a range of activities, including:

- Spatial relationships.
- Planned maintenance.
- Space planning.
- Asset systems analysis.
- Disaster planning and recovery.
- Post-occupation evaluation.
- Energy use and efficiency.

How BIM could be used

Good communication and levels of information will drive successful outcomes and the application of BIM can provide and enhance both of these requirements, and thereby provide an invaluable aid at all stages of the asset life cycle.

Where BIM has been used during the delivery phase of an asset, the information contained in the Project Information Models (PIMs) may be used to create the asset database, which in turn will form the basis of the Asset Information Model (AIM). Providing an as-built model (as part of the asset handover) provides a valuable resource for owners and facility managers to link operations, maintenance, and asset data required to effectively operate the facility (Figure 10.1).

The AIM also provides an accurate record of the completed space and can include links to all relevant facility information (for example serial codes, warranties, and the operation and maintenance history of all the components within the asset or facility). The AIM will utilise a common data environment (CDE)

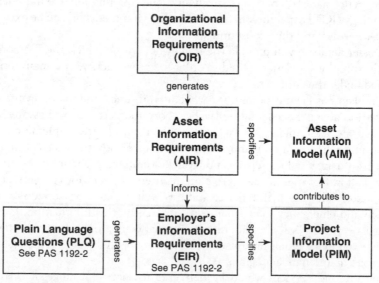

Figure 10.1 Relationship between elements of information management. PAS1192-3:2014

providing a 'single source of the truth' without having to maintain separate asset management systems.

Asset management systems are used to support financial decision-making, short-term and long-term planning, and maintenance scheduling. Using information in an AIM, facilities managers can: evaluate the cost implications of changing or upgrading building assets; track the use, performance, and maintenance of assets for the owner, maintenance and financial teams; and produce accurate schedules of current company assets for financial reporting and estimating the future costs of upgrades or replacements.

Impact on traditional process(es)

The following outlines some of the key areas where BIM may impact the delivery of FM services.

Spatial relationships

BIM can provide visualisation, access to locations and relationships of asset systems and equipment, especially those concealed in walls, floors or above ceilings without opening those walls, floors or ceilings.

Unlike 3D CAD files, BIM models can extend the information pool beyond geometric data. This may include weight, power supply, manufacturer, commissioning data, life expectancy and the like, along with warranty details and history of repairs. Through the use of mobile technology this information may then be assessed by a maintenance technician when inspecting and maintaining assets in the field.

BIM can provide Globally Unique Identifiers (GUID); these are machine-interpretable unique identifiers in the model, which can be used to link to other systems (including Building Management Systems (BMS), Energy Management Systems (EMS), Computerised Maintenance Management Systems (CMMS), Computer-Aided FM Systems (CAFM) along with other systems used to support FM). It can also support the creation of zones identifying areas serviced by common components, e.g. fire alarm or HVAC systems and capture system relationships, (e.g. which final distribution boards are served from, which sub mains distribution boards, switchgear and ultimately, which transformer).

Planned maintenance

BIM can assist with automating the creation of equipment inventory lists, populating facility management systems, such as a CMMS, and reducing maintenance of facility data for facility management activities.

Through the use of standard data transfer protocols, asset rich data may be migrated from the as constructed BIM model using common data formats, such as COBie or IFC. This reduces human error and the resource demands required to undertake the validation.

Space planning

Using BIM for space management enables the FM team to analyse the existing use of space, evaluate proposed changes, and effectively plan for future needs. Having accurate and detailed space information is especially useful for planning renovation projects, where all or part of an asset may need to remain operational during construction.

Asset systems analysis

Tracking performance data from the asset systems and comparing these values to design model predictions enables facilities managers to ensure that the asset is operating to specified design and sustainable standards, and to identify opportunities to review and/or modify operations to improve system performance.

Designers can also use this data to validate and refine their predictive models and evaluate the impact of proposed materials and system changes to improve performance.

Whilst Asset Systems Analysis typically focuses on mechanical systems and energy use, it can also include naturally ventilated facades, lighting analysis, airflow analysis using computational fluid dynamics (CFD), and solar analysis.

The predicted energy performance can be compared with actual performance data from the BMS sensors and meters. The examples in Figures 10.2 and 10.3 illustrate how software may be used to compare simulated energy consumption with actual consumption across the entire building.

This type of data and analysis would enable building operators to understand when and how energy use differs from predicted performance, providing a feedback loop of lessons learned and troubleshooting. During operations, it

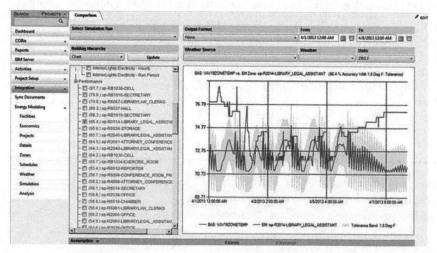

Figure 10.2 Screen shot showing predicted energy use vs actual energy use analysis.

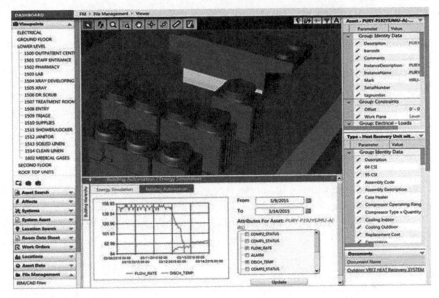

Figure 10.3 Screen shot showing BMS sensors values – properties/documents are shown on the right.

enables facilities managers to understand how the asset or facility was intended to be operated to achieve optimum performance. Feedback regarding the actual operation as compared to the design assumptions, would then allow more realistic and accurate energy predictions during future design phases, although the lack of published collateral suggests that this is still an area to be fully exploited.

Disaster planning and recovery

Using BIM to provide emergency services access to critical building information can improve the efficiency and effectiveness of their response and minimize the safety risks. Combining asset information, such as floor plans and equipment schematics, with the dynamic real-time state information provided by a Building Automation System (BAS) could provide emergency responders with valuable information to support better decision-making during crisis and disaster response. The AIM could be used to clearly display where the incident is located within the building or facility, possible routes to access, along with details of any hazards that first responders should be aware of.

Post-occupancy evaluation (POE)

POE is the process of obtaining feedback on performance in use and is typically carried out during the first three years of an asset's life. The value of POE is being increasingly recognised and it is becoming mandatory on many public projects. POE is valuable in all construction sectors, especially healthcare, education, offices, commercial and housing, where poor performance will impact on running costs, occupant well-being and business efficiency.

Combined with POE the concept of GSL also seeks to compare the required performance outcomes with actual performance outcomes.

The purpose of GSL and POE are to:

- Optimise the operating performance within the operational budget as soon as possible.
- Align the operating performance with the required performance outcomes set at the start of the design and construction period.

Typical facility operation is based on feedback from occupants allowing operators to address energy efficiency issues on the basis of complaints. The operators will usually adjust local set points instead of troubleshooting the problem and resolving it.

Energy use and efficiency

With the increasing demand for a more energy-efficient built environment, the construction industry is faced with the challenge of ensuring that the energy performance predicted during the design stage is achieved once a building asset is in use. However, there is significant evidence to suggest that assets are not performing as well as expected, and there is a significant gap between design stage target estimates and actual energy performance.

The traditional methodology tends to be a comparison of energy performance at a system level, as opposed to an asset level. For example, ventilation or air-conditioning systems are treated as single entities, whereas they are actually made up of a number of interconnected components, such as fans, pipework, cooling, heating source and terminal devices. The accuracy of the outputs of the energy model could be skewed by performance of individual components (for example, an oversized pump may compensate for errors caused by an undersized fan).

Traditional FM methods do not combine spatial and thermal perspectives, and they do not consider the relationships between components of energy systems.

To drive energy efficiency, one firstly needs to understand the performance of existing infrastructure and then to determine and advocate improvement initiatives over time. The capture of real-time information through the use of BIM could therefore be used to enhance the following:

- Metering requirements need to be articulated and provided as part of the construction project to enable all relevant energy consumption data to be collected around the facility or asset. Such metering will need to be by system, by floor and by department, to enable zonal, plant, service-level and operational data to be collected.
- Once metering is in place, energy data would need to be recorded

methodically at regular intervals in the right format to provide the inputs required to enable relevant analysis to be undertaken.

- The analysis of the energy usage and consumption data collected as the facility or asset is utilised, provides the real-time opportunity to not only understand how the building or facility is performing at various levels and by various energy sources, but also the opportunity to continually assess such performance against alternative options in terms of not only enhanced plant maintenance or service efficiency initiatives, but also plant renewal and/or specification enhancement decisions.
- Energy cost capture, as discussed above, will enable informed decisions to be made regarding the viability of energy-saving specification choices, maintenance systems and plant replacement options.
- Optimised metering, energy utilisation data gathering and cost capture will provide the information needed for energy efficiency assessments to be made and compared to alternative options, as the facility is maintained and used by the client. Day-to-day and periodic energy use assessments provide the management information needed to enable alternative energy source and plant specifications to be efficiently and effectively considered.

Once set up, assets may be tagged with the energy data outlined above, providing the energy management required to monitor usage from individual plant level to system, departmental, floor, operational/business function, public spaces and whole facility provision.

Tools capability

The scope, complexity and breadth of information required to support FM services makes BIM ideally suited to provide the necessary framework and depository for the collection, and the subsequent management and analysis, of relevant asset data for efficient FM delivery.

The following provides a summary of the tools typically employed to support FM, which can be incorporated into an AIM, ensuring the integrated capture and use of the information driving the FM service.

Computer-Aided Facilities Management (CAFM)

CAFM systems provide facilities for managing leases, space, resources, utilities, equipment and the like so that all of these are known and maintained in good order, can be monitored in terms of use, charged as necessary to particular departments, and purchased or disposed of according to an asset's operational use.

CAFM systems bring together disparate sets of data into a single integrated whole. For example, a space management system will utilise buildings, floors, areas, furniture, equipment and people for space analysis, reporting and chargeback.

Traditionally, CAFM systems have been populated by resurveying, rather than reusing the information collected during the delivery phase of an asset.

Computerised Maintenance Management System (CMMS)

The primary function of a CMMS is focussed on the maintenance of plant and equipment, and it typically contains information about the following:

- Scheduling routine work order tasks; planned preventative maintenance.
- Managing work order requests; reactive maintenance tasks and minor works requests.
- Recording asset history.
- Managing inventories, such as spare parts for plant/equipment and consumables like oil and grease.
- Assisting with statutory compliance.

Integrated Workplace Management System (IWMS)

An IWMS can be defined as a software platform that helps organisations optimise the use of workplace resources, including the management of real estate, infrastructure and facilities assets. Facilities such as office space, meeting rooms and workplaces may often appear to be well utilised. However, more detailed analysis can often show that utilisation is significantly under capacity. An IWMS provides the tools to measure and analyse space utilisation, identifying opportunities to increase the effective use and value, eliminate underperforming facilities and spaces, or change their functions.

An IWMS may incorporate:

- Real estate management.
- Capital project management.
- Sustainability and energy management.
- Maintenance management.

These systems might be further integrated with any of the following to enhance the level of FM service being provided:

- Geographic/Geo Spatial Information Systems (GIS).
- BMS.
- EMS.
- Enterprise Resource Planning (ERP) system. Broadly speaking, ERP refers to automation and integration of a company's core business.
- Life safety systems.
- Mobile construction field management software.

GIS

A GIS is capable of capturing, storing, analysing and displaying geographically referenced information. This is used to visualise, manage, analyse and collate

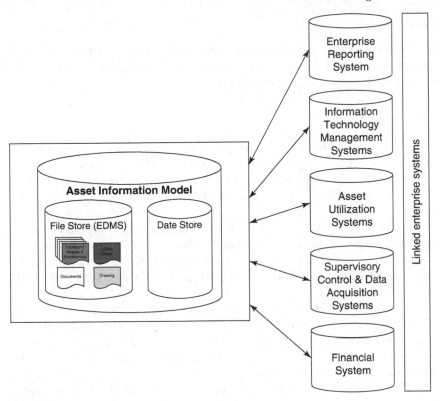

Figure 10.4 Interface between the AIM and existing enterprise systems. PAS1192-3:2014

data based on location. A GIS can be used by facility managers for space management, visualisation and planning, and emergency and disaster planning and response, as well as many other applications.

Energy Management Systems (EMS)

Energy management software may be incorporated into FM applications to allow real-time data flow between BIM and the energy simulation programmes to compare predicted and actual energy consumption. The software needs to be able to drill down to identify zones and components that may be deviating from acceptable limits, which could trigger a more detailed review by the FM team.

The use of integrated energy management (linked to the Building Management System (BMS) or Building Automation System (BAS)) may help the FM team to identify any problems associated with energy efficiency and resolve them in a structured, holistic and considered manner compared with traditional FM.

BMS/ BAS

A BMS or BAS is a computer-based control system that controls and monitors mechanical and electrical equipment such as ventilation, lighting, power systems, fire systems and security systems.

Case study details

This case study uses the specific example of an Private Finance Initiative (PFI) acute hospital (70,000m²) to illustrate the service profile to support provision of the hard FM services required to maintain and ensure operational performance in terms of use by NHS staff, patients (resident or visiting) and any other public or contracted visitors.

The case study also draws on the co-authors' experience in the preparation and agreement of cost estimates and commercial terms on a number of PFI contracts to support the delivery of Education and Government Accommodation, together with the management of life cycle and hard FM services during the operational phases of these projects.

A successful project will provide the required outputs for all stakeholders, achieving targets in finance, building and service performance which, in terms of the case study, were specifically related to patient care, patient experience and hospital availability, and where all stakeholders needed to be considered to ensure successful service provision.

Bearing in mind the service profile, timing, client and stakeholder expectations in terms of service provision, reactive maintenance and the constant availability of the hospital, BIM provided an ideal opportunity to collect, store, use and manage information to deliver clarity and completeness over the entire asset life cycle.

The hard FM service scope included the maintenance of the building elements classified in accordance with the Building Cost Information Service (BCIS) over the contract term.

Through the experience of the case study the following provides a few examples of how and where BIM might be used to support the delivery of hard FM services.

Asset information

Against each hospital asset collected in BCIS format, essential construction and facility information can be stored for reference and service management in accordance with PAS1192-3:2014.

Such information is likely to include the following:

• Particular specification details providing clarity on the initial installation. This is a critical first step in providing the asset details required to enable and support robust FM. These details will typically include information such as the asset descriptions, insulation characteristics, performance and output criteria. Supplier and installation contractor details can also be stored for ease of reference.

- Quantification of assets, providing the necessary scope of installation across the facility. This data can also be presented to not only quickly provide total quantities, but also the location of such assets across the facility, in as much detail as required.
- Against each asset, the scope of Planned Preventative Maintenance (PPM) can be added, providing the information needed to drive the timing and extent of technical service and resource requirements.
- Reactive Maintenance (RM) incidents can be recorded against each asset, including the extent of work carried out. This data might then be tracked and analysed to identify issues that may be assessed as either to be expected or flagged as warranty or defect related.
- Replacement requirements can be included, providing expected replacement intervals over the contract and an outline of the forecast scope. This data could be periodically reviewed to allow an up-to-date assessment of forecast replacement activities and costs to be communicated as required. Within a fully operational hospital such plans for major replacement works will be of great importance and relevance, not only to the hard FM provider, but also to the client and the impacted departments within the facility.
- Actual asset replacement activities can be recorded, proving a clear record of actual work carried out. This scope can be analysed to compare with previously forecast scope and also be used to inform and amend future forecast replacement works as necessary.
- Costs can be captured to the extent required, budgets can be allocated and costs can be recorded against each asset and then summarised to suit particular business requirements. A focus on budget allocation across assets in line with the building or asset element schedule as noted above, together with actual cost capture in the same format will provide ideal budget versus actual cost information for review and analysis to support a project by project comparison and also to inform the future programming of work.
- Health and safety compliance requirements, recorded and referenced to reflect legislative and business driven requirements to ensure compliance and safe working practices.
- Resource requirements can be identified by total hours, skill requirements, shift patterns, supervisory needs, management expertise and organisation. A resource plan will be developed from the data recorded against assets, providing the information needed to formulate the optimum level of management, supervisory, planning, and technical and trades staff with a level of confidence underpinned by the comprehensive capture of service requirements.
- Technical resilience within an operational facility needs to be identified and managed. Single points of failure need to be identified and strategies

to maintain facility operation need to be formulated and recorded against the relevant assets. Design should ideally take account of resilience issues allowing for a sufficient number of key plant items, such as boilers, to enable continued facility operation should plant or equipment need to be taken off-line for maintenance. If the design does not allow for this, alternative solutions need to be formulated, agreed and recorded against assets, to be utilised when maintenance and replacement works necessitate assets being removed from service.

- Technical system interfaces need to be highlighted against assets, providing information relating to the interdependence of systems and functionality, together with relevant commissioning data. This will be much-needed material in ensuring any maintenance issues are dealt with comprehensively by taking all relevant systems into consideration, and thereby ensuring the facility remains available for use.

The facility owner/operator will need to continue to maintain the AIM throughout the life of the asset to reflect changes to the facility.

Asset conditionality

In the case study example, the hospital needed to be maintained to the required standard and was subject to client checks and audits. The asset database and register held in the AIM can be used to store relevant condition information, which is likely to include the following:

- As condition assessments are made of the facility, asset by asset, the resultant findings can be stored for ease of reference, and any necessary works required flagged for remedial action. A condition grading system of 'A' (brand new) to 'D' (immediate replacement required) could easily be utilised and assets annotated to suit. Any remedial works highlighted can be scheduled to support work plans over any given period, based on forecast workload from the condition assessments carried out.
- In addition to asset condition, criticality assessment data will need to be held to enable the prioritisation of any work identified during the operation of the facility. A ranking system of '1' (insignificant to operations) to '5' (catastrophic to operations) – or other appropriate classification – can be utilised, making reference to asset importance a quick and meaningful process, and providing a logic to any such prioritisation in programming and budget allocation.
- Management outputs, including not only overall condition assessments, but also asset, function and system assessments may be derived. In addition, well-informed and soundly based plans of works from annual plans to five-year plans and beyond, may be formulated, to be managed alongside resources, works programmes, facility operation requirements and budgets.

Performance management

Using the hospital example, performance of the facility will need to be managed and the required performance measures can be stored against assets, providing the management data needed to track and manage output to ensure the facility is performing to the required service levels in the contract. This information needs to be readily available and will likely include the following:

- Trend logs can be generated and managed, listing performance monitoring and condition monitoring measurements to overlay against those expected by asset specifications. Such analysis plays a pivotal role in determining maintenance actions required to ensure asset and building or facility optimum performance in terms of fabric, services, environment and cleanliness.
- Plant downtime records may be kept against assets. Such data will identify workload effectiveness and asset reliability, and will be used to drive continuous improvement in terms of workload allocation and asset speculations.
- Service effectiveness can be measured and stored, measuring the mix of planned and unplanned/reactive work carried out to assets. This would provide management data for the identification of assets in need of replacement resulting from frequent failure and also for improvement of maintenance levels as appropriate.
- Asset mean time between failure records can also be kept and stored for the identification of assets in need of replacement, and also for improvement of maintenance levels as appropriate.
- Maintenance task completion times and closeout information can be stored and analysed to identify any significant variability along with asset maintainability issues, for attention and action.

Asset Management Plan

For the successful management of a facility, the mobilisation phase must be used to generate the asset management information listed and discussed above.

The culmination of this data gathering exercise is the generation of an Asset Management Plan in line with PAS 1192-3:2014, an example template for which is provided in Appendix 1.

BIM allows the required data to be stored at individual asset level for the facility, enabling real-time data to be gathered, managed and utilised, to ensure efficient and effective on-going maintenance management.

Issues/benefits

There are a number of issues to be considered and managed in relation to BIM for FM and these are outlined below.

Data will drive successful outcomes in the majority of cases, for example in the case of hard FM services for a hospital, where such data will include the extent and timing of maintenance required to ensure on-going functionality of plant and equipment in patient and non-patient areas and clinically critical and non-critical areas.

In addition to ensuring that all the relevant information for efficient FM delivery is available, such data also needs to be readily available, up to date and in a suitable format.

The data structure is therefore a key consideration. PAS1192-3:2014, BS1192-Part 4:2014 and RICS NRM3 provide guidance on recommended data structures to ensure accessible and manageable storage.

Early BIM adopters have tended to focus on design and construction, and as a result there is little empirical evidence on the implementation and the benefits achieved through the use of BIM during the operational phase of a facility.

BIM applications are evolving and standard formats to organize and share asset information are far from fully mature, and will be dependent on the adoption of a unified classification system.

Historically, construction data has not moved seamlessly into the FM world, and there are a number of possible reasons for this:

- Typically, as-built data has essentially comprised disparate sets of drawings and lists of parts. Such information can require considerable time to make it useful for FM.
- The fragmented and unstructured nature of the construction design information has often meant that the 'as built' information often ends up being in a very different form from that set out in the designs.
- Data is often incomplete or uncertain and the time taken to audit it, approached the time which might be taken to survey from scratch. For example, there may be a great deal of information about M&E schedules, however there may be a lack of useful information about maintainable assets, such as part numbers, warranties, maintenance schedules, etc.

As a result many believe it easier to resurvey the asset to gather the information required rather than sort through huge quantities of data.

The volume of data produced in the design and construction process is vast, and only a small part of this is of relevance to the FM team. As noted above, the development of a unified classification system should enable the relevant information to be more easily transferred into the FM systems.

The development of an AIM to identify, structure and store relevant asset data in such a way as to generate and enable efficient and effective FM will provide benefits to the project, the FM business and the client.

Improved asset information

The closer integration of AIM and facilities systems should result in improved asset management, reducing or eliminating time locating required asset information.

This will help owners and operators improve efficiency, make better-informed decisions about their assets, support improved decision making through the availability of rich information, and drive savings throughout the entire life cycle.

Streamlined handover and more effective use of data

A key benefit of integrating BIM with FM is that key data can be captured directly from the design and delivery phase and does not have to be re-entered or created downstream. This reduces the data entry cost and potentially generates improved quality data.

This in turn can result in smoother handover and the creation of required FM and asset management data. It also helps FM staff understand how to operate and maintain the asset, which is further supported by the use of GSL.

Asset operation

With the improved information available there are a number of benefits during the operational phase, for example:

- Availability of improved information when it is needed rather than maintenance technicians spending time looking up information from disparate sources – drawings, equipment documents (i.e. providing a 'single source of the truth').
- Reduced cost of utilities (energy and water) due to improved maintenance data, which supports better planned preventative maintenance planning and procedures. Plant and equipment will operate more efficiently when properly maintained.
- Reduction in equipment failures that result in emergency call-outs/repairs and disruption to occupants/users.
- Improved inventory management of spares and consumables.
- Tracking of asset and equipment history.
- Support for business continuity planning and disaster recovery scenarios.
- Location aware model of equipment, fixtures and furnishings, with supporting comprehensive data sets.
- Maximisation of asset life expectancies through extensive use of planned maintenance strategies.
- Display of real-time data.

To projects

Benefits for design and construction teams:

- Reduces costs of re-documenting 'as-built' drawings and information, and undertaking field surveys for building or facility renovation projects. Savings could occur from reduction in time to verify field conditions, change orders due to unforeseen conditions, reduction in destructive testing and repair costs.

- Greater accuracy in energy model assumptions and better estimation of energy performance.
- Design and construction teams can provide improved systems and facilities due to more informed equipment selection and specifications based on feedback from operations.
- Improved commissioning through understanding the impacts of individual Mechanical Electrical Public Health (MEP) services components on overall MEP systems. The classic example is adjusting air conditioning components to a particular space, which if undertaken in isolation may affect the remainder of the system because of the change in airflow. As adjustments are made to individual components, such as a terminal box, the overall system performance may be analysed and adjusted.

To the businesses

Benefits to the FM business will include improved business function definition:

- Certainty and clarity of contract obligations and performance.
- The ability to roll out systems and approaches to other sites.
- The compilation of business development case studies.
- The generation of empirical data for feedback into the contract and for future business development.

Benefits for building operators:

- FM costs can be inflated as the result of incomplete asset registers. An accurate equipment inventory can reduce FM costs by identifying and tracking facility equipment and the areas of the facility.
- As the FM system can be automatically populated it can reduce time creating equipment inventories rather than having to collate from disparate data sources – plans, specifications and submittals.
- An accurate asset register can generate energy savings by identifying all facility components that affect energy usage, require maintenance and assist in statutory compliance.
- Reduce risk and uncertainty of performing work task orders by identifying components that are not easily identified (e.g. valves).
- Maintains links to equipment histories, which can be fed back into condition survey assessments. An accurate asset register reduces the possibility of unexpected costs for unforeseen repairs by identifying accurate equipment locations and subcomponents.
- Optimise performance by comparing actual to predicted energy performance generated from BIM using energy simulation software.
- Through the integration of BIM, BAS, CMMS and GIS data, better interrogation and access to controls, schedules and readings will be possible. Cost and performance trending can be used to troubleshoot high or repeat

work order areas and identify customer satisfaction or building/facility performance issues.

- Reduce time by eliminating additional trips to the same location to carry out unscheduled work orders by providing accurate site conditions and maintenance information with improved coordination.
- Reduce cost of repairs by providing faster response times to emergency work orders.
- Mobile access to BIM using the appropriate hardware/software in the field allows access to related data/information without making repeat trips back to the office.
- The availability of real-time data is a key asset for the facilities manager, enabling early decision making and prompt action where necessary.
- The delivery of FM services will benefit from the integration of information generated by the various systems to provide more efficient management of assets.

Benefits for the asset owner should include visibility and engagement regarding service level assurance, energy efficiency, facility availability, building condition achievement and contract clarity, which should lead to greater efficiency and reduction in cost.

Benefits for occupants/asset users:

- Increased satisfaction as a result of reduction and/or quicker resolution of unscheduled work orders.
- Reduce unscheduled work orders and increased communication between tenants and building maintenance workers regarding scheduled work orders.

The use of BIM to support FM offers the potential to reduce the operational expenditure for asset owners. BIM should also require less effort during the early stages of a project's life to determine the materials and systems that will require optimal maintenance management throughout the life of the assets.

Through the integration of information systems such as EMS, BMS, CAFM or CMMS and geospatial software, benefits may include:

- Operational expenditure reduction for asset owners, as the traditional task of identifying and collecting information on the materials and systems required to provide optimal maintenance management throughout the operational life of the asset post-handover should be reduced or eliminated.
- Efficiencies in the life cycle process through not having to validate 'as-built' drawings and information and not having to undertake field surveys for building refurbishment projects.
- Building performance optimisation by comparing actual and predicted energy performance generated from BIM using energy simulation software. Further efficiencies can be realised when using this in conjunction with feedback information from the BMS.

- Reduction in cost of repairs by providing faster response times; this is facilitated by 'on the job' access via the use of mobile technologies to asset information combined with 3D visualisation and possibly augmented reality to locate the asset. This is especially useful for plant and equipment located above ceilings or services concealed within the building's fabric.

In addition to the above, there are other benefits linked to the management, organisation, infrastructure and strategy of organisations, such as improved workforce productivity through better facilities operation.

Measurement of benefits

As already stated, there is very little collateral on actual efficiency savings achieved in the operations phase through the use of integrated FM and BIM.

Benefits may include a range of efficiency improvements, including consistent asset condition, consistent facility availability, high customer satisfaction, positive operational feedback, application to other projects and bids, continuous improvement and lessons learnt by all stakeholders.

Evaluation of return on investment (ROI) should not only focus on the operational improvements generated by BIM, but those of a managerial, organisational, infrastructure and strategic nature.

It is likely that the most significant benefits of BIM for an asset owner are those of an organisational and strategic nature, which are difficult to measure using quantitative means.

Summary/commentary

The process and experience from the estimation of [PFI] FM costs and life cycle costs, to the operational management of FM services and life cycle management services, has enabled the industry to move from theory to practice, and to gain a high level of knowledge as to how well-structured asset data can improve operational management, a key facet of which is the strong link between the construction phase and the operational phase of a facility in terms of design, specification and quantification.

FM encompasses the delivery of strategic and operational objectives for a facility and its stakeholders, including the facility users, the client and the FM company. FM provision includes hard FM building fabric and plant services and soft FM such as cleaning, catering and laundry services.

The effective delivery of FM services requires real-time data, in relation to both technical specifications and building performance, to be structured sufficiently for such data to be meaningful, understandable and obtainable, where and when required. An ideal data structure, building on the link from construction to operation, is the utilisation of the BCIS elemental estimation structure, encompassing all elements of a building.

The example of the need to effectively mobilise the operational delivery service for a hospital has identified a clear need to identify, obtain, structure and effectively retain and make accessible key asset information to ensure that the knowledge of the facility is such to enable a successful transition from the completion of construction to the commencement of service operations. The generation of an asset database and asset register containing such information within a resilient ICT system is critical. BIM provides the opportunity and ability to link data in a BIM model to a database of building assets and their relevant data records.

Such a database storage facility will provide the platform for building or facility system analysis, along with resource planning for the facility. Data stored will cover a range of asset information including specifications, quantities, costs, preventative maintenance tasks and outcomes, reactive maintenance tasks and outcomes, design resilience, health and safety information and guidance, condition and performance indicators, energy utilisation and efficiency, soft FM service interfaces, and will enable the preparation, development and real-time upkeep of a fully informed Asset Management Plan for the facility.

BIM provides the vehicle for design, construction and FM data to be held and linked together in one system, designed to suit the needs of particular projects, clients and key stakeholders. The benefits of this to the project, the FM business and the client will include improved asset information and visibility. It will mean a smooth handover from construction to operation in the case of FM service mobilisation, and a strong, meaningful and applied link between design, construction and operation and the provision of efficient and effective FM services to users and clients.

To date, BIM has focused on the design and construction of the built asset. On the operational side, BIM can be used to improve the performance and productivity of a facility owner's construction, operation and maintenance processes.

Project teams can capture information required to manage and operate a facility from BIM used through the project delivery phase.

The AIM provides a database of assets to assist in efficiently maintaining and operating the facility, so that end users will have the information they need to operate the facility through a CDE without having to maintain separate asset management systems.

As the industry matures, more data will become available to encourage investment in BIM to support the operational phase.

For the larger construction companies who are likely to have already invested in BIM the decision to extend the investment into the operations phase should be relatively straightforward, especially as the cost (and potential savings) of operating the asset could be as much as four times the cost of designing and constructing the facility (i.e. 80 per cent of the WLC) over a typical 30-year life.

Whilst BIM for FM may still be at an early stage of maturity, the potential to transform the way facilities are managed, along with the opportunity to inform the design and construction process, where a small change at design stage may have a significant impact (cost, time, quality) further down the 'life' of the asset, should provide the necessary incentive for the QS to develop their service offering. An example here is providing adequate consideration on how the asset will be accessed for maintenance and operational purposes; poorly accessible plant and equipment will tend to receive less maintenance which in turn can result in reduced performance, higher energy use and poor customer perception.

It is perhaps worth noting the range of disparate systems and the opportunity that this may present to the QS in consolidating FM data and the inherent benefits of BIM in allowing the reuse of common data.

Appendix 1: Example asset information plan template

- Introduction
  - Facility Background
  - Facilities Operated and Managed
  - Services Provided
  - Objectives
- Levels of Service
  - Service Level Requirements
  - Legislative Requirements
  - Response Times
  - Rectification Times
  - Forward Service Demands
- Review Process
  - Review Systems and Frequencies
  - Formats and Requirements
  - KPI and Audit Systems
  - Service Specifications
  - Condition Standards
  - EHS Policy
- Trend Analysis
  - Environmental Management
  - Maintenance Levels and Management
  - Life Cycle Management Systems
  - Condition Assessments
  - Benchmarking and Best Practice
  - Workload Review and Management
- Performance Measurement
  - Service Specification Measurement
  - Service Monitoring and Review Systems
  - Asset Data Analysis
  - Service Performance Reporting
- Improvement
  - Gap Analysis
  - Management Strategies
  - Periodic Reviews
  - Efficiency Initiatives

- Management Systems
  - PPM Management Systems
  - RM Management Systems
  - Emergency Management Systems
  - Information Management Systems
  - EHS Systems
  - Company Policies and Procedures
  - Procurement Systems
  - Accountancy/Finance Systems
- Life Cycle Management
  - Design Specifications
  - Planning/Work Identification
  - Condition Assessment
  - Replacement Plans
  - Life Cycle Policies and Procedures
  - Life Cycle Optimisation
  - Risk Management
- Work Plans
  - Daily Maintenance Plans
  - Weekly Maintenance Plans
  - Monthly Maintenance Plans
  - Annual Maintenance Plans
  - Annual Life Cycle Plans
  - Five Yearly Life Cycle Plans
  - PPM Programmes

References

BSI (2013) *PAS1192-2 Specification for information management for the capital/delivery phase of construction projects using building information modelling*, British Standards Institute, London.

BSI (2014) *PAS1192-3 Specification for information management for the operational phase of assets using building information modelling*, British Standards Institute, London.

BSI (2014) *BS1192 Part 4 Collaborative production of information: Fulfilling employer's information exchange requirements using COBie – Code of practice*, British Standards Institute, London.

BSRIA BG 4 (2009) *The Soft Landings Framework, Report BG 4/2009*, Building Services Research and Information Association, Bracknell, UK.

Cabinet Office (2011) 'Government Construction Strategy', HMSO, London, Online. Available   HTTP: <http://www.cabinetoffice.gov.uk/resource-library/government-construction-strategy>.

HM Government (2013) 'Construction 2025 Industrial Strategy: Government and Industry in Partnership' BIS/13/955, London, Online. Available at <https://www.gov.uk/government/publications/construction-2025-strategy>.

RICS (2014) *New Rules of Measurement 3: Order of cost estimating and cost planning for building maintenance works, First Edition*, RICS, London.

11 Dispute resolution

Andrew R. Atkinson and Christopher Wright

Introduction

In the absence of any known or reported disputes arising on the early round of BIM compliant projects, this chapter assesses the likely impact on the range of services currently offered and provided around dispute resolution.

Author biographies

Andrew R. Atkinson

Andrew is a quantity surveyor with a background in consultancy and public service. He lectures at London South Bank University in contract administration, project management and quantity surveying. He has acted as principal and co-investigator for several publicly funded research projects covering subjects from housing adaptations for people with disabilities, knowledge management, buildability, human error and safety. He publishes regularly in academic journals, and is author of the JCT Contract Administration Pocket Book (Routledge). To ground academic work in current practice, Andrew maintains a small surveying consultancy.

Christopher Wright

Christopher founded Christopher Wright & Co LLP, Solicitors, a law firm specialising in construction law. During his career he has worked 'in house' in both public and private sectors, and in private practice. He advises on a range of construction contracts, both domestic and international, and deals with disputes. His work also includes the construction aspects of process plants. He is a practising construction adjudicator, accredited by TeCSA (the Technology and Construction Court Solicitors' Association) and a CEDR accredited mediator. His publications consist of articles and a text book on housing improvement law.

Service profile

Dispute resolution is a broad term covering all means of settling disagreements between contracting parties by the intervention of an external agent. Dispute resolution services are divided into two approaches; consensual and binding, within which there are several variants.

Consensual resolution

In consensual resolution the parties consent to work together, with external assistance, towards resolving a dispute. Accordingly, it is sometimes called assisted negotiation and methods include:

1.   Mediation.
2.   Conciliation.
3.   Neutral evaluation.

Binding resolution

In binding resolution the parties are unwilling to work together and resolution must be by a third party. Methods include:

1.   Expert Determination.
2.   Adjudication.
3.   Arbitration.
4.   Litigation.

All forms of dispute resolution outlined above involve a similar method of working. The essence of the process is that an external referee is asked to come to a decision on the basis of material to which he or she has had no access and on the basis of representations from people that he or she does not know. Whether an English advocacy process or a more inquisitorial process is used by the referee, the individual must very quickly achieve mastery in the minutiae of the case. With advocacy, material is presented by opposing parties with a view to supporting their case and the referee must cut through representations in order to come to a decision. With statutory adjudication (the most common forum for dispute resolution in UK construction) this must be done within 28 days of referral and with very little scope to extend this time.

Additionally, the process may be complicated by the parties 'mixing and matching' resolution methods. For example they can agree, in the same building contract, to disputes on measurement and valuation being resolved by the expert determination of a quantity surveyor. Other disputed matters, such as delay or technical performance, could be dealt before a specialist Arbitrator.

Dispute prevention

Dispute prevention mechanisms can often be employed in larger building and civil engineering projects, where it is more common for disputes to arise. Early warning notice procedures have been built into the contractual machinery of a number of contracts over the years. For example many JCT forms have included provisions requiring the service of notices of delay and loss and expense. More recently, especially on larger projects, this approach has been adopted more widely than for the prescribed and fairly narrow range of circumstances covered in some standard forms. For example, NEC3 (Institution of Civil Engineers 2013) requires the issue of an early warning notice to be followed by a site meeting at which the parties try to amicably work out a solution to the problem.

The concept of trying to head off formal disputes has been developed further on some projects by the practice of appointing, in advance, an individual or panel of individuals to hear disputes as and when they arise. Prevention mechanisms allow early pre-emptive resolution of disputes before the parties become entrenched. This approach also ensures that agreed resolution machinery is in place before a dispute has arisen – a much easier proposition than trying to gain agreement once the parties are already involved in a dispute.

In the absence of any meaningful case history relating to BIM Level 2-compliant projects, this chapter will firstly consider how the use of BIM might either prevent disputes or assist the dispute resolution. It will then go on to consider a case study for a project pre-dating the introduction of BIM to assess how the use of BIM might have influenced the outcome.

How BIM could be used

Clearly, one of the intended outcomes of BIM is to prevent disputes arising. However, assuming disputes will continue to occur despite its introduction, BIM will more likely influence the conduct and outcome of dispute resolution in three areas:

Transparency

BIM has the potential to remove some of the need for representations in dispute resolution. One of the under-emphasised characteristics of the sharing of a common data environment (CDE) is that it is possible to invite parties not involved in the original model construction 'to the party' later in the design-construction-operate-demolition sequence. This could include the disputes referee, who would have a clear record of who did what and when.

One of the largest expenses in any litigation and many construction arbitrations is the cost of making disclosure. This is the process whereby each party is obliged to list all material documents in its possession (plus those which were in its possession, but which it no longer has) and make them available to the other party. Although the courts are increasingly moving towards a regime of e-disclosure,[1] whereby the parties exchange electronic data for use in a hearing, the very transparency of BIM means that it will be much more difficult for parties to make incomplete disclosure, whether accidentally or on purpose, as this will be open to view by the other party through the CDE. Many delaying arguments about disclosure, which would otherwise have to be determined by the referee, can be avoided. This should streamline, but not remove the need for advocacy in summarising and presenting material to the referee, whether that is done on a 'live' basis, or simply in the form of preparing documentary submissions.

Audit trails

Related to transparency is the clear indication in the published standards and protocols[2] that audit trails will be maintained within BIM as a product of change control. The audit trail will be available to the referee in the form of data timeline snapshots allowing comparison of information at different times. These snapshots will help in explaining why changes were made and how decisions to make changes were arrived at. Change has led to catastrophe as well as financial failure in the past. Knowing the perpetrator of a change, its rationale and timing should greatly assist the analysis of the merits of a case. Transparency and audit trails should therefore empower disputes referees to assess the cause, liability and value of a dispute.

Embedded decisions

There will also be the potential to embed decisions in the model. Defective physical elements in a model might remain in order to provide an audit trail of changes, but the case history of a referee's decision and reasoning could be embedded in the corrected version.

Tools

Achieving the benefits of transparency, audit trails and embedded decisions will require the identification of assistive software. The particular problems, discussed further below, relate to the complexity of BIM and the need, in many instances, to come to decisions quickly. To the extent that these are not catered for in existing modelling tools, the following information presentation tools may be necessary:

Examination of decisions – who, what, when, where and the how of key decisions related to configuration, specification, cost and time during each stage of the project and including brief development reflected in the Employers Information Requirements (EIRs) along with any subsequent design/construction changes.
Simplification and extraction of information – some techniques for removing extraneous information, or extracting only the information related to disputed elements.
Mapping – PAS 1192-2:2013 provides clear guidance on mapping the key roles, responsibilities and decision points throughout the project delivery cycle.

Issues/benefits

The implementation of virtual object-oriented model-based information representing buildings or other assets in place of more traditional two-dimensional drawings, specifications, bar charts and cost plans presents problems both for dispute resolution practice and for the disputes coming before a referee.

Implications for dispute resolution practice

Implications for practice can be summarised in that the referee, whether judge, arbitrator, adjudicator or mediator will be required to understand and practically interface with the BIM environment. Judges, although specialising to an extent through the courts within which they practice (e.g. in the Technology and Construction Court – TCC), typically rely on outside expertise to explain technical issues. However, court cases form a small minority of disputes. They have a large legal content and it is appropriate for judges to concentrate on the law and accept expert advice. Arbitrators and adjudicators are expected to be both knowledgeable in the technology under consideration and in the law. Indeed, it is for this dual knowledge that they are often selected. Effective dispute resolution for a BIM compliant project would demand that the referee speaks 'the language of BIM' as well as having a good knowledge of law and the technical matters under consideration. Specifically, the referee is faced with:

- Complexity.
- Prescriptive documentation.
- Access issues.
- Dual agendas.

Complexity

Complexity is evidenced in the various publications emanating from such bodies as the British Standards Institution (BSI) (2013, 2014) and the Construction Industry Council (CIC) (2013a). These documents introduce complex terminology, which is almost entirely separate from the traditional lexicon of construction communications. The lack of mapping of one terminology on another is possibly advantageous in that it encourages a wholly new way of thinking, but it taxes the intellectual capability of the technically qualified, let alone of those qualified in non-cognate disciplines such as law.

Prescriptive documentation

The interpretation of BIM has become highly prescriptive, with the division of its implementation into rigid levels of detail, formally defined in the CIC BIM Protocol along with the relevant BS and PAS standards. An example illustrates the point:

> The EIRs shall be consistent with other appointment and contract documents in use on the project, which in turn should be aligned with industry standards such as the RIBA Plan of Work or APM Project Stages (PAS1192-2:2013, 9)

'Shall' in the above statement indicates a requirement of the PAS (referred to above) and 'should' a recommendation (BSI 2013, iv). Thus it is a prescriptive requirement of the PAS that the EIRs (Employer's Information Requirements) are consistent with the contract documents (for example, the Employer's Requirements in a JCT Design and Build form of contract), but it is only a recommendation that the EIRs and contract documents are aligned with divisions of work by stages as published by the RIBA or the Association of Project Management (APM). Leaving aside the possible need for the referee to adjudicate on whether the PAS had been complied with in a particular instance, the prescriptive nature of the requirement means that the referee will need a very thorough knowledge of the documentation in order to come to a sound decision.[3]

Access issues

Access issues relate to whether the benefit of transparency can be achieved in a dispute within a BIM project. Specifically, the CIC BIM Protocol carefully controls the use of information by defining 'permitted purposes'. A permitted purpose is:

> a purpose related to the Project (or the construction, operation and maintenance of the Project) which is consistent with the applicable Level of Detail of the relevant Model (including a Model forming part of a Federated Model) and the purpose for which the relevant Model was prepared.

(CIC 2013a)

Dispute resolution is not specifically included as one of these purposes and it could be argued that it is not related sufficiently to the project to be intended as a permitted purpose. This is because parties do not ordinarily enter into a contract with the intention of committing a breach giving rise to a dispute. The CIC BIM Protocol also makes provision for project team members to grant licences to the employer, but these are also tied to the permitted purposes, and guidance in the CIC BIM Protocol indicates that should further licences be required by the employer, then separate agreements will be needed from *each project team member*. If a referee is not considered a project team member and dispute resolution is not a permitted purpose, the referee could be in a worse position with regard to access to information than with a traditional paper-based project. At least for the latter, physical drawings and bills of quantities might be available for perusal!

Dual agendas

Dual agendas are apparent in the CIC BIM Protocol and in some reviewed publications, including Azhar (2011), Succar (2009), Gu and London (2010) and CMAA (2012). The expressed agenda is enabling the construction of an information model. The implied agenda is that of *collaborative working*. In some publications, BIM is put forward as a process enabling collaborative working, where designs are iteratively worked out between designer and constructors to a point at which they can be built. The problem with this dual agenda is that it implies a method of working (despite the CIC BIM Protocol expressly stating that it *'makes the minimum changes necessary to the pre-existing contractual arrangements on construction projects'* (iv) – author's italics).

In making decisions for disputes related to traditional projects, where full designs are handed over to contractors at the tender stage, referees will need to interpret BIM protocols liberally and in the context of the size and type of project.

Implications for disputes

The creation of new knowledge domains, procedures and roles in the move towards a BIM-enabled world will involve a similar change in the subject matter of construction disputes and a referee in a BIM Level 2 compliant project may be required to decide on some novel issues specifically engendered by the advent of a virtual environment (CDE). These could include:

Whether a protocol has been provided and followed

The staged provision of drawn, written and spoken information for a construction project has, traditionally, been managed fairly loosely. For example, in earlier versions of the JCT contracts, drawings were to be provided by the architect at a time, neither too close nor too distant from when they were required. In response to a case where a contractor asked for all drawings in short order to suit an ambitious internal programme (Glenlion v the Guinness Trust 1987), this laxity was tightened by amending these contracts to require architects to provide schedules indicating when drawn information would be available. The advent of BIM will tighten this further by driving amendments to standard forms requiring compliance with whatever BIM protocol is in use for the project. A referee will be asked to decide on whether a protocol has been incorporated in the contract and, if so, whether the parties have complied with it.

Failure to appoint a competent information manager

It is common with initiatives in the building industry for new roles (in this case of 'information manager') to be created with the employer being responsible for making the appointment. As with the principal designer (formally CDM

Coordinator under CDM 2007 in relation to health and safety procedures), this role may be taken by another named professional such as the architect, but there could be a separate appointment or the role could be handed down through the design-construction sequence as the project design develops. Should the information manager fail to act in accordance with the relevant protocol or task specification (for example, CIC (2013b) *Outline Scope of Services for the Role of Information Management*) one or other party to the construction contract may wish to take action. In the case of a contractor taking action, a referee might be asked if the employer is in breach of contract for failure to appoint a competent individual. Should the appointment be another named consultant or made on the advice of a consultant, the employer may well in turn look to that individual for redress through the consultant's conditions of engagement.

Failure to comply with the information manager's requirements

The function of 'information management' is set out in detail in PAS1192-2:2013 (Table 2, page 19) as a separate task. This includes activities to:

- Enable reliable information exchange through a CDE.
- Maintain and receive information into the Information Model.
- Enable integration and co-ordination of information within Information Model.
- Configure information for Project Outputs.
- Populate the information exchange format for the Information Model.

The '*Outline Scope of Services for the role of Information Management*' is further defined in CIC (2013b) as including:

- Establish a CDE.
- Establish, agree and implement the information structure.
- Receive information into the Information Model.
- Maintain the Information Model.
- Manage CDE processes and procedures.

In order to execute most of these activities, information and collaboration is required from the project team members. Where this is not forthcoming, the information manager is faced with trying to achieve compliance without any direct contractual tools. This has the potential to draw the employer into disputes over enforcement – the implication being that the employer will have to have a much more direct engagement in this aspect of practice. It is interesting that the problem largely arises because information management is separated out in the prescriptive documentation at the same time as all other 'project team members' are being drawn together in a spirit of collaborative working.

Whether collaborative working has been forthcoming from the project team members

Collaborative working is a major theme of BIM, and Choat and Steensma (2014) consider this theme in relation to the project team members' duties to warn of deficiencies brought to their attention. This general contractual duty is likely to become much more prominent in a transparent environment of a CDE. Access to a CDE gives opportunities for a party to peruse and comment on technical aspects tangential to their own activities. Disputes could arise over whether that party should have warned that a particular detail was defective or inadequate.

Similarly, Choat and Steensma (2014) suggest that there may be scope for a tortious duty of care being owed by one project team member (PTM) to another outside the contract. The duty of care is created by the project team member publishing models or sub-models in the CDE in the knowledge that they could be relied upon by other members in developing their own models. Again, a referee may be called upon to decide this matter, with a further complication that, as this relates to matters beyond express contractual provisions, it may not come within the scope of statutory adjudication and could involve referral directly to the Courts.

Whether a 'soft landing' has been achieved

Soft landings are defined in BSI (2013) as:

> a graduated handover of a built asset from the design and construction team to the operation and maintenance team to allow structured familiarisation of systems and components and fine tuning of controls and other building management systems (p51)

Whilst this may suggest a certain 'fluidity' around the precise nature of how soft landings will be achieved, BSRIA (2009) indicate that it will involve an extended rectification period (where the design/construction team is given an express opportunity to put right defects for an extended number of years rather than the normal year), there is likely to be increasing difficulty in distinguishing latent defects and 'snagging' items from wear, tear, misuse and vandalism. It may therefore be the case that disputes over the origin of post-occupation defects will increase in volume as a direct result of an increased opportunity to pursue them and will engage referees well beyond the end of the construction phase of the project. Additionally, contracts will need to be modified if it is intended to allow existing resolution mechanisms such as disputes boards or adjudicators to operate well into the occupation period. It is possible to avoid post-construction latent defects arguments by using project insurance and this is considered further below.

A case study

A brief case study will assist in illustrating how dispute resolution may be influenced by BIM. Whilst there may be little evidence to assess the impact of BIM compliant projects on dispute resolution, it is possible to look at a dispute where BIM was not used and to consider how it could have influenced the course of proceedings and resulting decision.

The case – background and award

The case in question involved the refurbishment of a substantial Victorian office building in central London. Included in the work was the insertion of a lift, together with associated alterations to the structure, provision of lift guides, supporting structural steel members and enclosing walls. Unfortunately, shortly after completing the work, the lifts began to experience problems evidenced by cracking of the masonry enclosing walls and loosening of padstones, upon which the supporting steel rested. This resulted in the lifts vibrating alarmingly and becoming unsafe to use. The occupying sub-tenant called in a structural engineer, who pointed out the defects and gave a view on their cause – *the encasing walls were insufficiently substantial to resist lateral loads imposed by the action of the lifts*. The sub-tenant naturally enough sought redress and the lease and contract structure allowed this.

Neither the principal tenant, nor the sub-tenant was the direct client under the contract. However, the principal tenant had the benefit of a collateral warranty with the main contractor. The lease terms between principal tenant and sub-tenant contained a provision requiring the principal tenant to pursue the sub-tenant's claims against the contractor through the collateral warranty. The construction contract was based on the traditional JCT98 private with quantities form (Joint Contracts Tribunal 1998), but with a contractor designed portion for the lifts and all builder's work in connection with the lifts. As is normal with traditional contracts of this type, the employer directly engaged a consultant design team, including an architect, structural engineer and quantity surveyor.

As the dispute was a matter relating to the original building contract, the parties used adjudication in accordance with the Construction Act[4] for its resolution. However, the adjudicator was not asked to make a monetary award, but simply to determine the contractual liability of the parties, presumably on the basis that the parties' respective quantity surveyors could then agree a financial settlement. Expert engineers, including the structural engineer acting for the sub-tenant, and a structural engineer acting for the respondent contractor, made representations on the cause of failure and suggested suitable remedial measures. These could subsequently determine the extent of damages should the contractor be found liable. Although there was some disagreement on detail, in general the engineers were agreed on the cause of the defects and remedial work required. The dispute was limited to determining whether the contractor was liable for the design and construction of defective walls. The sub-tenant and referring party (the principal tenant) contended that it was and the contractor disputed this, contending in contrast that the encasing walls to the lifts were not part of the lift installation or associated builder's work.

The contractor was relying on the fact that details (as subsequently amended) of construction for the encasing walls were provided in drawings produced by the employer's consultant structural engineers. These details put the walls outside the contractor's designed portion. The details were initially that the encasing walls were to be of 140mm dense concrete blockwork. Later they changed, first to what appeared to be lightweight dry-lined construction and finally to the 100mm dense concrete blockwork used in construction. Underlying the changes were considerations of limited space for constructing the walls.

Normally with a traditional JCT98 contract, a contractor is required to comply with designs provided by consultant architects and engineers employed by the client, but if the required work is part of a contractor designed portion, the design is provided by the contractor and the consultant engineer's designs are given for information only (perhaps in order to indicate the general positioning and size of elements and give suggestions for possible construction materials). In this case, the consultant's designs were quite specific and detailed the thickness and material for the walls. Given that it was common ground that the walls enclosing the lifts had failed and that remedial work was required, the main consideration was *whether the walls were part of the lift installation and 'associated builder's work' and therefore formed part of the contractor's designed portion.* If this was the case, the contractor was responsible for either using due skill and care in designing and constructing the walls, or for checking the consultant's design, before adopting it.

The adjudicator found for the referring party. The walls as part of the lift shaft were, on the evidence of the documents, within the contractor's design portion. *The contractor was in breach of the terms of the contract in that it had failed to use due skill and care in the design of the walls and had produced some defective workmanship.* This was despite the fact that detailed drawings were provided by the consultant structural engineer. The provision of drawings did not dilute the contractor's design obligation with regard to the design of the lift shafts and the contractor's design portion contract documents (in the Employer's Requirements) expressly stated that the contractor was to be fully responsible for the drawings for the shaft and pits. Further, the Employer's Requirements also stated that the contractor should become an integral member of the design team and to work with the architect, structural engineer and any other design team members in the development of the design and finalisation of systems and configurations. It also expressly provided that the contractor should be co-operative and pro-active in its design approach.

The contractor claimed that the Employer's Requirements did not contain fundamental performance information needed to design the lift shaft, but the adjudicator determined that if that was its view, the contractor should have complied with express terms of the contract to warn of discrepancies and divergences in documentation. It should also have warned of any inadequacy in the designs put forward by the structural engineer. The adjudicator decided that the contractor had adopted the design, but this did not absolve it of responsibility for design, nor for any subsequent changes required in later instructions. The adjudicator determined that a *'properly qualified and competent contractor experienced in designing and carrying out work of a similar scope, nature and size to the Works'* (the standard of care imposed by the contract) would, as part of their design obligation, either have carried out a design check itself, or warned the employer of the risk of proceeding without ensuring that a check had been carried out.

One question that was expressly NOT considered by the adjudicator, as it was outside his jurisdiction, was the possibility that the employer's structural engineer owed a general duty of care to the contractor, and that this duty was breached. The structural engineer, advising the contractor, considered that there were errors in the design of the lift shaft walls as constructed. If the employer's consultants knew that the contractor would rely on those designs, coming as they did from competent qualified engineers, then there might be a duty of care in tort. This possibility is more likely in a contract (such as JCT98) where the employer's engineer (through the instructing architect) has executive authority to issue instructions for the majority of the structural work. It is only in the exceptional circumstance related to contractor's design portions where this does not apply. This issue did not arise in the adjudication, but might be relevant in more collaborative scenarios adopting a BIM approach and how this may have impacted dispute resolution is now considered below.

How could BIM affect the progress of the dispute?

This dispute illustrates many problems that BIM should address. Despite the evident soundness of the adjudicator's decision given the terms of the contract and the facts, one cannot help having some sympathy for the contractor. With traditional contracts, such as the JCT98 'With Quantities' form, there is an expectation that designs will be produced by competent consultants acting for the employer and contractors will be responsible for executing them. If a consulting engineer provides quite specific design information regarding an element of construction, the contractor will often be tempted to comply without much regard to the fact that (for that particular element) the engineer does not have the authority to impose the design, or that the design is inadequate. The procedures and protocols of BIM should fill some of the gaps in responsibility and action revealed by the case in the following areas:

Divisions of responsibility

Both BSI (2013) and CIC (2013a) carefully define responsibilities for tasks. The division of responsibility for execution and checking of designs for the lift shaft walls would, therefore have been defined earlier in a BIM compliant project by the Master Information Delivery Plan (MIDP) and Model Production and delivery table (MPDT) forming Appendix 1 of the CIC BIM Protocol, and in compliance with PAS 1192-2:2013. The fact that parametric model elements for lift shaft walls are provided as outline for completion by the contractor would have been clear at the stage the contractor was being asked for input. Should input not be provided by the contractor, this would remain as a flag to be corrected, capable of follow-up by the information manager.

Clash avoidance

One of the drivers of the dispute was the conflict between technical requirements for the lift shaft walls and space restrictions. It is likely that the latter drove modifications to the extent that technical requirements could not be easily met. The adoption of BIM will actively control such clashes through inherent design coordination, thus providing early pre-emptive warning that the problem requires attention. It is therefore likely that the technical requirements would have been met and the dispute avoided.

Change control

Similar to clash prevention, change control with BIM involves changes to models. The effect on technical capability of the lift shaft construction would have been flagged by changes from 140mm blocks to framed dry lining and then to 100mm blocks. Had the contractor proposed an alternative form of construction, this would have been subject to similar change control and the adequacy of the proposal would have been checked.

Incorporation of contract requirements

Despite the fact that this was a project with a contractor designed portion, the managerial responsibility is well documented and the information delivery cycle in PAS1192-2:2013 would make it clear that there were differing responsibilities for (for example) the design of the general building structure and the design of the lift shaft walls.

Status control

Linked to a more rigorous division of responsibilities provided through BIM compliance is the more careful control of the state of information contained in the model. Models are required to be labelled for various purposes and failure to 'sign off' a design by an appropriate authorising authority (for example the consultant engineer) as 'for construction' would be flagged, thus avoiding the situation that sufficient technical input (such as engineering calculations for the walls) was missing for an element.[5]

Centralised responsibility – information management

Tying together the various information control measures above is the new role of information manager as envisaged by BSI (2013, 2014), and the Construction

Industry Council (2013a, 2013b). The appointment of this manager does not remove the need for decision making on technical information by an appropriate authority. However, it does provide a centralised driver for information. When combined with more systematic division of responsibilities, clash BIM/design coordination, change and status control, having single point responsibility to collect information on these aspects will ensure that each is considered.

It is, perhaps, also worth considering the cultural impact of the collaborative working inherent in compliance with BIM Level 2 project delivery.

How could BIM affect the liability of parties to the dispute?

A duty to warn?

A relevant question is whether the collaborative environment of BIM will see an expansion of claims based on a duty to warn. In the case study, the adjudicator determined that a duty was owed by the contractor under an express contractual obligation to cooperate with the consultant design team. It required the contractor either to check the wall loadings itself, or if it did not have that expertise, to warn the employer's design team to check the loadings before proceeding further. The risk, which it failed to manage, was that the loadings provided by the lift subcontractor were not checked by the employer's design team, and this fact should have been obvious to the contractor.

It can be difficult to imply a duty to warn into a contract, because generally the law works on the principle that where the parties have had an opportunity to put their obligations into a contract, it is not for Judges, or others to plug the gaps. If such a duty was to be imposed, the contract would say so. However, it is also possible that aside from an express contractual obligation, there is a duty to warn under the law of tort, in particular in negligence. As mentioned above, the duty in tort was expressly not considered in the case study, because adjudication is restricted to disputes based on contract.

A duty of care in providing design information?

In the case study, the contractor discharging its duty to warn might have made a difference to the decision, however, in the more visible arena of a CDE, many more parties will have access to specialist models and will be drawn to rely on them in order to complete their own models. Choat and Steensma (2014), as noted above, suggest that there may be scope for a tortious duty of care being owed, outside contractual arrangements, by one project team member to another in providing design information. If that is correct, the contractor in the case study might have been more easily able to seek a remedy from third parties – for example from the consultant structural engineer if its design was deficient, or from the architect for not following up the unapproved loadings. Future disputes (particularly where BIM is grafted onto conventional procurement systems without modification) might well see an increase in the number of disputes based loosely on 'lack of collaborative working', but which involve an increased general duty of care between contributors to the CDE.

Integrated project delivery models and project insurance

An increased duty of care in producing models or in warning of deficiencies therein, could be mitigated by exclusions of liability. However, if the contractor and design consultants exclude liability, one of the key advantages of BIM is lost, namely its ability to enable a much more collaborative approach than conventional contracts through the CDE. This problem is recognised in the USA, where 'integrated project delivery' is proposed to manage the design and construction team at the same time as BIM is adopted (CMAA 2012). Project team members agree to waive their rights to take action against fellow team members in exchange for a share in any overall gain made by the team. Members also agree to share losses.

In a similar approach, the UK Government is promoting 'Integrated Project Insurance' (Integrated Project Initiatives 2014), one of three new models of procurement incorporated within the *Government Construction Strategy* (Cabinet Office (UK) 2011). With BIM in mind, Integrated Project Insurance anticipates that all contributors to a construction project will combine to form an 'alliance'. Insurers have agreed to provide project-based insurance covering financial losses over and above an agreed pain/gain limit and including all the alliance members. The alliance members, including the client, agree that they will waive their rights to claim against each other and the project insurers agree to waive subrogation rights against all the alliance members.

It is, therefore, clear that, had the case study been a fully developed collaborative BIM project modelled on an Integrated Project Insurance alliance, the insurance policy may have covered the sub-tenants loss and the referral would have been unnecessary.

Could BIM have assisted the adjudicator in the case study?

Analysis of the facts outlined above, suggests that BIM probably would not have assisted the adjudicator in reaching the decision. That said, provided the adjudicator had the requisite knowledge of modelling to be able to access information, and provided that dispute resolution was a 'permitted purpose' for the use of model information, there may have been some marginal efficiency gains in presenting the case for the referring and responding parties.

However, the decision did not involve the analysis of large amounts of data and was more concerned with the meaning of such terms as 'lift shaft', 'lift well' and 'all necessary builder's work' – common phrases in the industry, but subject to varying interpretation. For other cases, where decisions are based on complex and voluminous interlinked information, BIM is likely to be a more useful tool to the referee.

Summary/commentary

New demands on referees

BIM should reduce the incidence of disputes by improving the consistency, quality and transparency of project information. When disputes do arise BIM also has the potential to make resolution easier. Information will exist in an integrated transparent CDE accessible to all project team members for all permitted purposes. The CDE is also expected to have an archive section expressly for inactive or superseded information available to others for 'contractual purposes or "discovery"' (BSI, 2013 page 44). Thus, it is intended that transparency and an audit trail will be available to those responsible for settling differences. For the sake of clarity, it would be a small amendment to protocols to expressly include dispute resolution as a 'permitted use' of all models and thus, remove arguments related to the 'disclosure of documents'.

In addition to the features of transparency and audit, it would not be difficult to embed dispute decisions in the model for archive and information purposes to be used by occupiers and on future projects. This could be extended, with suitable safeguards to provide a decisions database available at a technical level to designers and constructors, but also at a legal level of use to referees.

The biggest problems associated with dispute resolution in a BIM compliant project are the expertise required of the referee, the time available for a decision and the cost of relevant software – if separate proprietary software licences are required for access, although in many cases this may simply require the use of freely available viewer applications. Most referrals are currently to statutory adjudication, with very tight time limits and an expectation that the adjudicator will be able to come to a decision largely on the basis of his or her own knowledge of law and technology. That requirement will remain, but will be supplemented by a need for expertise in BIM and related technologies used on the project. As mentioned above, this expertise will need to be detailed and specific. Atkinson and Wright (2014) report that adjudication fees are relatively inexpensive, but the additional expertise requirements are likely to increase these substantially.

The case study involved a fairly conventional adjudication, where the adjudicator was a solicitor, appointed because of his legal expertise and ability to accurately determine the wording of the contract. Adjudicators also come from a wide range of other disciplines, including architecture, engineering, quantity surveying or accountancy. They combine adjudication with other professional practice, operating from a variety of sizes of organisation, from sole principal to major corporations.

Possible solution – disputes boards

The complexity of BIM indicates that some form of panel of 'BIM Enabled Referees' could be necessary in order to lessen the slope of the BIM learning curve and to avoid resort to paper documents when all else is handled electronically.

The referees should be appointed as disputes advisors, or disputes advisory boards to act over the course of the project. It is already becoming more common for parties to decide whom they wish to determine their disputes in advance, both at adjudication and at arbitration, and it would be a fairly small step, in a BIM setting, for the parties to agree from the outset to appoint an individual as a disputes advisor or a panel of individuals as a disputes advisory board.[6]

Other than the benefit of having BIM-capable referees available over the course of the project, there are several other advantages to disputes advisory boards:

- Matters can be referred for an opinion or a discrete finding on a particular point before the parties have gone too far down the wrong road. They can make regular visits to the project if required, in order to assist this process.
- Advisors can be appointed from different disciplines with, for example, an engineer for design issues, a quantity surveyor for valuation issues and a lawyer for interpretation. They could act individually depending on the nature of the dispute, or collectively in more complex situations.
- The same board can be appointed for separate contracts on different parts of the project thus ensuring consistency of approach and findings.
- It is also possible to empower a board to make a binding decision on an interim or final basis.

A problem with disputes advisory boards is the cost of the appointment. Giving access to a BIM model as it develops should be cheaper than expecting advisors to come to terms with a fully developed model at a later referral, but there will be a front-loaded cost in making the appointment. In the context of an expected reduction in the number of disputes as a result of more efficient information flow with BIM, there is a fine balance between incurring the front-loaded cost of a disputes advisor and leaving appointments to later. The weight of costs and benefits is likely to be different for each project and, on balance, it is likely that early appointment of a board, particularly for larger projects, would seem financially worthwhile.

The cost and value of dispute resolution with BIM

Costs of dispute resolution on BIM-enabled projects are, initially, likely to be higher. Training adjudicators and arbitrators is both expensive in itself, and will restrict the pool of qualified referees, also increasing costs. There may also be a cost associated with accessing information and, where disputes advisors or boards are to be used, an upfront cost of set-up and retention.

However, BIM should not increase the cost of the resolution process and may, in fact, slightly reduce it. Rather than the referring party producing hard copies of documentation as its exhibit to witness statements, the party could simply provide index references and leave it to the referee to access information electronically.

Once the initial costs of training, access and set-up are absorbed, savings should outweigh continuing costs. When combined with the benefits of more consistent and reliable decisions embedded in final models for posterity, the cost/benefit of the process should be strongly tilted towards value which, in turn, should also lead to an increase in service value – albeit these services may be more aligned to prevention rather than the resolution of disputes.

Notes

1   See Ministry of Justice (UK) (2014)
2   See for example PAS1192-2:2013 and Construction Industry Council (CIC) BIM Protocol (2013a)
3   It is surprising that the authors of the PAS have used 'should', when they could have used 'may' and put their meaning beyond doubt. When interpreting any words in a document the test as to what the word means is an objective one (Chartbrook Limited v Persimmon Holdings Limited [2009] AC 1101) and the word 'should' has often, although not universally, been found to be mandatory in meaning by the Courts (R v Board of Trade, ex p. St Martin's Preserving Co Ltd [1965]1QB 603). Plainly the authors of the PAS do not intend that particular consequence.
4   Housing, Grants, Construction and Regeneration Act 1996 Part 2 as amended by the Local Democracy, Economic Development and Construction Act 2009 Part 8
5   Relying on status control does, however, have its problems whilst BIM practice is developing. Wall (2014) cites a litigation case where an architect failed to communicate the order of construction essential to fit all components into a building, relying instead on BIM to do this. The consequences lead to a substantial out of court settlement and a timely reminder of the dangers of overestimating the capabilities of immature BIM.
6   A number of terms are in use for the various types of board which do roughly the same job.

References

Atkinson, A. R. and Wright, C. J. (2014) 'Containing the cost of complex adjudications', *ASCE Journal of Legal Affairs and Dispute Resolution in Engineering and Construction (JLADR)* – accepted for publication.
Azhar, S. (2011) 'Building Information Modeling (BIM): Trends, benefits, risks, and challenges for the AEC industry', *Leadership and Management in Engineering*, 241–252, July.
BSI (2013) *PAS1192-2:2013, Specification for information management for the capital/delivery phase of construction projects using building information modelling*, BSI Standards Limited, London, UK.
BSI (2014) *PAS1192-3:2014, Specification for information management for the operational phase of assets using building information modelling*, BSI Standards Limited, London, UK.
BSRIA BG 4 (2009) *The Soft Landings Framework*, Report BG 4/2009, Building Services Research and Information Association, Bracknell, UK.
Cabinet Office (UK) (2011) *Government Construction Strategy 2011*, Cabinet Office, London, UK.
Cabinet Office (UK) (2014) *New Models of Construction Procurement*, Cabinet Office, London, UK.
CIC (2013a) *BIM Protocol*, Construction Industry Council, London, UK.

CIC (2013b) *Outline Scope of Services for the Role of Information Management* – CIC/INF MAN/S: 2013, Construction Industry Council, London, UK.

Choat, R. and Steensma, A. (2014) 'Liabilities for design defects in a collaborative, integrated digital age: The state of BIM in the UK: the route to Level 2', Unpublished paper.

CMAA (2012) *Managing Integrated Project Delivery*, Construction Management Association of America, McLean, USA.

Gu, N. and London, K. (2010) 'Understanding and facilitating BIM adoption in the AEC industry', *Automation in Construction* 19, 988–999.

Health and Safety Executive (2015) *Managing health and safety in construction: Construction (Design and Management) Regulations*, HSE Books, Sudbury, UK.

Higgin, G. and Jessop, N. (1965) *Communications in the Building Industry*, Tavistock Publications, London, UK.

Housing, Grants, Construction and Regeneration Act 1996 Part 2 as amended by the Local Democracy, Economic Development and Construction Act 2009 Part 8.

Institution of Civil Engineers (ICE) (2013) *NEC3 Engineering and Construction Contract* (clause 16), Thomas Telford, London, UK.

Integrated Project Initiatives (2014) *The Integrated Project Insurance (IPI) Model; Project Procurement and Delivery Guidance*, Integrated Project Initiatives, Cabinet Office, London, UK.

Joint Contracts Tribunal (1998) 'Standard form of Building Contract, Private, with Quantities 2003 amendments', Sweet and Maxwell, London, UK.

Ministry of Justice (UK) (2014) 'Civil Procedure Rules and Practice Directions, Part 31, Practice Direction 31B – Disclosure of Electronic Documents', Online. Available HTTP: <http://www.justice.gov.uk/courts/procedure-rules/civil/rules/part31/pd_part31b#1.1)>, accessed 23rd April 2014.

Succar, B. (2009) 'Building information modelling framework: a research and delivery foundation for industry stakeholders', *Automation in Construction* 18, 357–375.

Wall, G. (2014) 'BIM – Fully embrace it with your eyes wide open', asgard project solutions, Online. Available HTTP: <http://asgardprojectsolutions.cmail1.com/t/ViewEmail/i/62EC166756EC3922>.

Cases

Chartbrook Limited ᵛ Persimmon Holdings Limited [2009] AC 1101)
R v Board of Trade, ex p. St Martin's Preserving Co Ltd [1965]1QB 603)
Glenlion Construction Limited v Guiness Trust [1987] 39BLR 89

© SUE PITTARD

12 Summary/conclusion

Generally

This final chapter considers what BIM means for the quantity surveyor (QS), reflecting on the case studies featured to assess the opportunities and threats it poses. It considers how the QS role may be developed to exploit the use of BIM, and particularly the impact it may have on traditional services, as well as considering some of the new services which may evolve through the availability of richer data sets, along with the likely impact on skills and, in turn, the knock-on effect for education and training.

BIM is largely about the data and this is evident through most of the case studies featured. Data collected and structured for re-use throughout an asset's life cycle will enable a much richer blend of analyses to deliver a broader range of support services through the design, construction and operational phases of future projects. Better-quality information will foster better quality services to allow more informed decision making. The QS will therefore need to become familiar with the standards and protocols along with the technology required to read, interpret and analyse this data.

Traditionally, the level of approximation and informed 'guesswork', particularly in the early stages of a project life cycle, has been high. No matter how experienced, it is still guesswork. Key decisions are often based on judgement rather than hard data, which leads to an increase in the level of risk and uncertainty. BIM offers the ability to fill this information gap and reduce the level of guesswork. Improving the quality of data/information will reduce risk, and as a consequence, lead to a potential increase in value. Ultimately, all projects are about risk and value, and both should continue to be at the hub of the QS service portfolio. There is no doubt that BIM will foster innovation across the industry as businesses look to exploit the data produced, both in terms of enhancing their own business and to benefit their clients. For the QS, this is likely to manifest itself in the enhancement of existing services along with the development of new service streams to provide a richer, broader service portfolio. The context within which these services can be developed and delivered around BIM are discussed below.

Information

As the name would suggest, BIM is all about information. For too long the value of data and information has not been understood. In truth, historically collecting and collating data and information was an expensive and time consuming exercise. It still can be, but that is where the value of BIM becomes evident. The cost to the industry of rekeying data is no small matter, not only in terms of hard cash, but also in lost productivity. To liberate this value and increase productivity there are two key aspects which need to be addressed; the *quality* of the data/information and its *reusability*.

The quality of the data and information produced by the stakeholders to a project, or the owners of an asset, is their individual responsibility. In a BIM-enabled environment, because the information is reused and will be relied on by others – possibly for many years, it is therefore important to all involved that the information is accurate, and this accuracy is the responsibility of the information provider – which leads on to the difficult problems of *who* checks information and *who* warrants that it is correct? Will this lead to the insurers taking an interest and developing products to handle this perceived risk?

Will the checking and warranting of the information put into the model become a part of an Information Management role, or is there another 'new' role centred round the checking and warranting of information?

The reusability of the data/information will be determined by its accessibility and classification. With cloud computing and ubiquitous internet connectivity, accessibility can now almost be seen as a given, which just leaves the knotty problem of classification, although this is now being addressed through initiatives such as the NBS BIM Toolkit and Uniclass 2015.

As discussed in Chapter 6, the role of Information Management, and the management of information, is pivotal in a BIM-enabled project. Whilst this may be a function shared across the design and delivery teams, it is likely that there will be a lead appointment, and this is, indeed, embodied in the CIC Outline Scope of Services for Information Management and mandated in the CIC BIM Protocol. The role has three principle components:

- Managing the common data environment.
- Project information management.
- Collaborative working, information exchange and project team management.

Some projects have engaged the services of a specialist to perform some parts or all of this function. However, as the importance and significance of this role becomes clearer it is likely that responsibility for this role will sit with those already involved in the design and delivery, and the QS is well placed to lead this function – either as an extension to the normal appointment, or as a specialisation. The QS will, more often than not, be the first discipline to deal with *all* of the project information, and is well versed with checking and auditing information in terms of its compliance and appropriateness.

As BIM-enabled projects become the norm, the next challenge will not be the lack of sufficient data or information, but that there will be *too much*, and hence the core QS skills of reading, interpreting and analysing data will be a valuable commodity.

Measurement

Of all the traditional QS services, measurement is probably the one single function which is likely to see the most significant change as the result of BIM.

Historically, measurement has been a rather mechanical, heavily labour-intensive task. The scale rule and dimension paper, alien to most outside the realms of the QS, have been the main tools of the trade for decades. The introduction of computing, particularly during the 1980s, transformed this task, allowing the direct entry of items and their dimensional values, along with semi-automated billing. This process was further improved with the use of CAD, firstly 2D and subsequently 3D, replacing paper drawings and enabling automated measurement through new tools.

The use of paper drawings is certainly now at a premium, and probably only exists (for measurement) where the QS has them printed or plotted for their own use. The electronic drawing has, for all intents and purposes, replaced paper drawings. However, in truth, this transition to digital drawings probably had more to do with reducing the cost of printing at source (i.e. architect or design engineer), than as a means of fostering more collaborative working!

However, those using CAD-based tools soon realised that automated measurement relied on the use of common standards, which has been the Achilles heel to achieving the promised potential, productivity and efficiency gains, and preventing the realisation of the true potential of digital measurement beyond on-screen digitising.

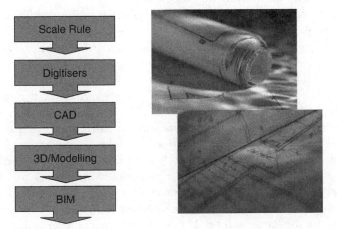

Figure 12.1 Measurement evolution.

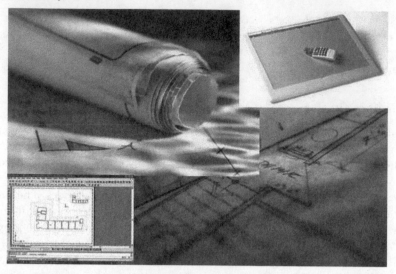

Figure 12.2 Digitising.

Along the way, the QS has also flirted with digitisers (a sort of electronic drawing board – see Figure 12.2 below), which have all since been made redundant by the world of CAD – and more latterly, BIM.

The ability to quantify a detailed accurate model is nothing new. The issue has always been the lack of standardisation and classification. However, as the industry adopts and adapts to the new standards, assisted through tools such as the NBS BIM Toolkit, automated measurement will become a more common reality, reducing and perhaps even eliminating the need for traditional measurement skills. This trend is also likely to be further fuelled by the increase in off-site fabrication and on-site assembly – Design for Manufacture and Assembly (DfMA) – resulting in the need to measure and control production and assembly, rather than traditional construction methods.

The QS has to learn to live in a world where quantification will be something that is undertaken by a machine. QSs should not hanker back to the 'good old days', but rather grasp the nettle of the technology and develop the high-value analytical and strategic services that will be required to fully exploit BIM, as exemplified through the many examples featured in this book.

Are the RICS New Rules of Measurement (NRM) suite BIM compliant? As all three parts pre-date the new digital tools for BIM, it is likely that there will need to be a period of review and some modification to reflect feedback and use. The key issue will be in the form of classification and coding, which may well result in some further work to ensure compatibility and compliance, and this will require the RICS to work closely with the other professional institutions to achieve an *industry*-created solution to an *industry*-created problem.

Despite all of the above, the ability to measure is likely to continue as a core skill, at least for the foreseeable future, therefore the temptation to discard it from education and training requirements too quickly should be resisted.

Technology

There is no doubt that technology will continue to develop, both in terms of computing power and in its potential to impact and influence both our working and social behaviour. The challenge will be how the industry continues to harness and develop technology in parallel with its capacity to adapt and absorb it for everyday use.

Perhaps one of the criticisms of BIM has been that it may have become overly complex, creating some unnecessary barriers to more widespread adoption. This is acknowledged by the UK Government in its Digital Built Britain (DBB) strategy published in 2015:

> A reoccurring theme with much engineering and especially BIM software has been the fact that it attempts to solve all user functions and in doing so becomes so complex only a small community of interested parties can actually use the systems. Normal lay users are mostly conversant with applications such as email and social media, both of which perform complex processes, yet manage to present the user with clear simple interfaces. Our aim must be to present the day to day user with useful easy to consume and interact with information and knowledge.
>
> HMG (2015: 25)

It will therefore be important to look at how technology can be made more accessible so that it might gain wider traction and acceptance. Indeed, the DBB strategy references the uptake of social media tools and suggests that the industry should learn from them when developing appropriate tools to encourage wide adoption and usage (HMG 2015). The DBB strategy goes on to say:

> We want to make fully computerised construction the norm and ensure that the benefits of these technologies are felt across the UK and support the export of these technologies and the services based on them.
>
> HMG (2015: 5)

This aim will only be possible once more widespread engagement throughout all areas of the construction industry is achieved, so that the use of BIM technology becomes 'business as usual', i.e. where technology and working with technology is second nature (HMG, 2015). For the QS this will mean not only harnessing the use of technology to support traditional services, but also, and perhaps more importantly, to leverage the use of new technologies in parallel with innovation and the development of new service streams required to deliver added value throughout the asset delivery life cycle. If the QS requires any motivation to embrace the digital economy then they need look no further than the DBB strategy, which states:

> As construction transforms itself into a modern digitally enabled industry, it will need fewer quantity surveyors and bricklayers and more people with

qualifications in production management, logistics, supply chain management, collaborative systems and data analysis. HMG (2015: 29)

BIM may therefore actually drive the need (and opportunity) for two levels of technology sophistication; the specialists who create and develop these technologies and services, and the construction and asset management generalists who use the technologies (HMG 2015). The QS should find opportunities at both levels.

More recent developments have seen the use of immersive technologies, which will probably do more to encourage the use of BIM for FM – as clients, such as Crossrail, see the benefits this can offer.

Opportunities and threats

Does BIM provide an opportunity or threat to the QS? As indicated earlier in the book, it rather depends on whether your 'glass is half empty or half full'. There will be those who see BIM as a threat to the established order of things and will try and resist the change that it brings. Certainly there is no doubt that BIM will bring change, and this is evident through many of the case studies featured in this book – and for those unwilling to adapt, then BIM most definitely does pose a threat. However, for those willing to adapt and to develop their skills beyond the traditional boundaries, BIM offers many opportunities and scope to expand the traditional QS service portfolio. What form some of these new services may take is considered later in this chapter. It is, however, equally important that the industry does not get carried away by the euphoria of change. If the result or output doesn't look right then it probably isn't! Garbage in, garbage out (GIGO) is still as relevant today as it was with the introduction of computing back in the 1970s. All that has changed is that these systems have become much faster, more powerful, more accessible and have the ability to produce even more 'garbage' if left unchecked! Quantity surveyors should, therefore, resist the temptation to forget or discard core skills (baby out with the bath water syndrome). BIM doesn't replace the need for expertise and will place greater emphasis on, and need for data analysis.

What opportunities and threats does BIM pose? It certainly offers the potential to reduce, and even eliminate the need (and burden) of some time consuming tasks such as measurement/quantification. Whether you see this as a threat or an opportunity to make better use of your time (and skills) again depends on your ability (and willingness) to adapt. Table 12.1 indicates some of the points to be considered when assessing the impact of BIM:

Added value

What value does BIM add? Certainly, there is a view to suggest that clients do not expect to be charged additional fees for the use of BIM, or for the role of information management. This is very much viewed as an 'in the box' service, and

Table 12.1 SWOT analysis

Strengths	Weaknesses
• Reduced time/costs	• Technology incompatibilities
• Ease of audit	• Perception of high investment cost
• Shared information	• Complacency
• Reducing errors and inconsistencies	• Education and training
• Provides centralised management entity	• May encourage too much detail
• Confidence in information	• Human error, parties not following
• Finding information easier	process ie. lack of discipline
• Less time consuming ie. eliminate some	• Learning new systems
manual tasks	• Luddite view/barrier to change
• Reinforces professional skills and values	
• Reduces risk	

Opportunities	Threats
• Creates working modules for asset lifecycle	• Insecure data storage - hacking/stealing data
• Creates a centralised database, time saving	• Refusal to adopt
• To improve/refine supply chains	• Reduce value of services
• Easier to project manage	• Ownership issues, legal obligations for professionals
• Increased value of data collected to support business	• Blurs professional boundaries
• Increased scope of services	• Job security / certainty impacting motivation
• Freeing up time for more valued services ie. reducing need for traditional skills	• Reduces/impacts service portfolio
• BIM Manager / coordinator	• Increased competition for QS service space
• Improve cashflow, earlier fee payments	• Cost to implement
• Optioneering & related value	• Automates core service ie. quantity measurement – reduced value
• Increased value of new skills and services	

therefore something clients believe they were already paying for. Indeed, to some extent, BIM has increased the level of fee competition, putting pressure on fees for many of the more traditional QS services, reflecting the increased efficiencies now expected from BIM. As a result, there has been a move to increase the value of these services by providing, for example, proactive and dynamic optioneering along with improved cost risk identification and mitigation, in order to maintain fee levels, i.e. charging the same but delivering more (added) value. Others have looked to develop new services around the availability of a more complete and reliable master data set, such as services around performance measurement and management, as featured in Chapter 9. Furthermore, with the increasing demand for a more energy-efficient built environment, the construction industry is faced with the challenge to ensure that the energy performance predicted during the design stage is achieved once a building asset is in use. With many assets not performing as well as expected and often suffering a significant gap between design-stage target estimates and actual energy performance, there should be further opportunities for the QS to develop and apply their measurement and analytical skills.

These enhanced services do not always mean doing more work, in fact BIM should mean doing less (time consuming) work for more fee – a sort of more for less!

Measuring the benefits of BIM

Whilst there is very little hard data on the benefit that BIM may bring to a project, portfolio or programme, there does appear to be plenty of anecdotal evidence that BIM brings benefits, both tangible and intangible.

If the industry is to achieve the Government's target of a 20 per cent reduction in the capital cost of buildings and a 33 per cent reduction in whole life cost – as embodied in both its 2011 construction strategy, and more recently within its Construction 2025 report respectively – then perversely, this will need to be measured. The QS will therefore need to understand where the savings are being made and where efficiencies and productivity can yield further cost savings in the design, construction and operation processes. Indeed, the creation and availability of consistent historical life cycle cost data for analysis and benchmarking is also likely to see many cost related services increase in value as the industry seeks to respond to the UK Government targets, specifically for the 33 per cent reduction in whole life costs.

Cost reduction is only one aspect of the benefits that BIM can bring to projects and the industry at large. There also needs to be an understanding of the *value* which BIM brings to projects, i.e. better design solutions because of the increased collaboration between consultants, safer sites because of improved methods of working and better understanding of the risks. BIM presents opportunities to get involved earlier and throughout the asset life cycle. It also offers a real opportunity to engage more directly with their clients to improve business outcomes, rather than the traditional focus on construction cost management.

Table 12.2 below indicates a simple methodology for exploring the possible benefits that can or could flow from the use of BIM.

Table 12.2 Examples of possible benefits

Potential benefits	Can be quantified and valued	Can be quantified, difficult to value	Identifiable not quantifiable
Freed staff time	✓		
More work taken on/won	✓		
Fewer mistakes		✓	
Better quality output		✓	
Greater job satisfaction			✓
More creative thinking time			✓

Adapted from Walshe and Daffern (1990).

Measurement will therefore depend on the nature and type of benefit, albeit the many intangible benefits, both in terms of the project and a client's business, may prove difficult to quantify. However, the opportunity and motivation to extend the core QS skills of measurement and analysis beyond traditional construction cost management should prove sufficiently attractive and rewarding, and particularly during the formative phase of BIM adoption where evidence to demonstrate benefits to all stakeholders is likely to determine the rate of implementation.

The future QS

It appears that much of the current debate about the future of the QS is around how BIM can be used to improve and develop traditional services offered by QSs, with very little on the potential and the opportunities that BIM brings to develop new service streams. An indication, perhaps, that the profession has yet to fully understand the impact (and implications) of BIM, and to recognise the opportunities (and threats) it presents. This may be an area where the RICS can support its members.

Collaboration

Probably the single biggest ingredient and requirement for BIM is collaborative working – something for which the construction industry isn't particularly renowned. Collaborative working will, therefore, require a major cultural change which isn't going to happen overnight (or without some effort). The UK construction industry is founded on entrenched (often outdated) working practices, and so the introduction of new standards and protocols will not be enough on their own. As the saying goes, you can lead a horse to water, but you can't make it drink! BIM requires a major shift in mindset. Indeed, this has prompted the Edge report, *Collaboration for Change* (Morrell 2015), which explores the key issues facing professionals and their institutions, citing collaboration as *the* catalyst for change.

In addition to a change in attitude, BIM also requires strategies to support the creation and sharing of information within the common data environment (CDE). Typically this involves allocating spatial ownership and responsibility of data – essentially subdividing the project to allow more than one stakeholder to work on the project models simultaneously. An example of how this might be applied to tunnel design is shown in Figure 12.3 below.

Collaboration also presents some interesting challenges to the delivery of education and training, not only for those entering the industry, but perhaps more importantly, for those already in it! Collaboration may present the biggest challenge to the future role of the QS, not that the QS doesn't collaborate with the other members of the project team, but that someone has to lead the move to more robust and practical collaboration. The QS is ideally placed to lead this development in the industry.

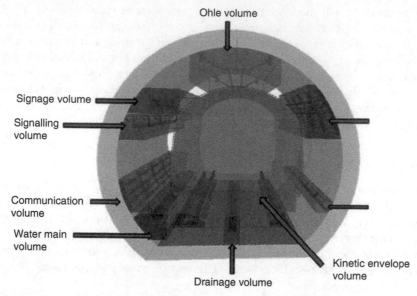

Figure 12.3 Volumes within a tunnel design for spatial co-ordination.
PAS1192-2:2013

What are these new services?

Not surprisingly the majority of the new services are likely to be based around the collation, analysis and reporting of data and information. This book has outlined a number of the new services:

- Information manager/management.
- Master data gatekeeper.
- FM/Asset management.
- Performance benchmarking.
- Risk management (develop new techniques around BIM).

There are also a number of additional services, which could be developed by the QS related to the data and information that will be readily available from the BIM environment, such as:

- Project/Programme delivery manager/coordination.
- Production management.
- Embodied carbon (services around assessing and managing).
- Client relationship services.

Education and training

The advent of BIM has enormous repercussions for education and training. As BIM is probably more revolution than evolution, it presents some radical changes to the way QSs do what they do, and this means learning new skills and new ways of working.

BIM will have a significant effect on higher education courses and the accreditation which is required of these as a foundation for professional qualifications. Indeed, BIM has already prompted a review of the RICS pathway guides governing the routes to professional qualification, which has in turn resulted in a review of the corresponding degree programmes, with the need to blend the new skills required to work in a BIM environment with the core skills to meet accreditation. This may drive some to challenge how much time they devote to the historically fundamental core skills of measurement and taking-off, as the industry becomes more familiar and confident with analysis and quantification through BIM.

There has also been a move towards assessing and recognising those with the particular roles and skills that will be needed in order to successfully deliver BIM in the industry. Indeed the RICS is the first professional institution to offer BIM Manager certification.

The growth and availability of BIM training courses offered by commercial providers has perhaps suffered from a lack of regulation, which has created some confusion and uncertainty. This is being addressed by the BIM Task Group through publication of the BIM Learning Outcomes Framework intended to encourage more consistency and standardisation. Time will undoubtedly tell whether this achieves the intended uniformity necessary to support universal understanding and interpretation.

Whilst education and training will play an undeniably pivotal role in achieving the required industry transformation, the need for separate or specific BIM-related courses is probably just a measure of maturity, and that longer term, as BIM matures to 'business as usual', this will form part of the underlying learning and assessment in the same way that information technology (IT) has become just a part of our everyday life.

....and finally

Whilst the editors had planned to use real examples to illustrate how BIM can be used to support the various services provided by the QS, the lack of available case study material is perhaps a reflection of a lag in maturity (to that expected), which is also echoed in the NBS BIM Survey (2015) – see Figure 12.4 below.

What is the reason for the dip in 2014? Is this the dawning realisation that using BIM is more than just the use of 3D technology and that there is much to do to fully implement BIM within individual projects and the industry as a whole, or is it that the hype is now disappearing?

Much has been written and said about collaboration, and there is no doubt that collaboration is at the heart of BIM. If the benefits and efficiencies of BIM are to be realised all of the stakeholders to a project need to work towards a common goal; that of delivering the client's requirements on time and at the right cost. True collaboration between parties with conflicting goals is very

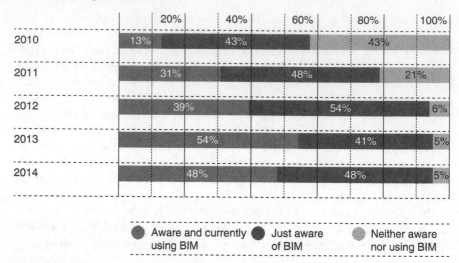

Figure 12.4 BIM awareness and usage.
NBS (2015)

difficult if not impossible to achieve, and much needs to be done to explore and trial how all of the parties to a project can be incentivised to work towards a common goal.

BIM is not about technology, it is not about 3D design or dynamic fly-throughs. BIM is about the *data* and *information* and the relationships that underpin the design, construction and operation of built environment assets. Implemented properly, BIM will revolutionise the construction industry and all of those who are involved. This book ends where it began, in its simplest form, the outcome of Building Information Modelling (BIM) is nothing more than delivering facilities or assets more efficiently – doing what we said we would do, in the way we said we would do it and by the time we said we would deliver it!

Action plan for BIM:

- Engage with other project stakeholders.
- Understand the information requirements of others and how you can improve.
- Identify and plan for the necessary education and training skills.
- Awareness and familiarisation with the required standards and protocols.
- Understand the requirements of BIM Level 2 compliance – if there isn't any EIR documentation then request it!
- Identify internal BIM champions to lead.
- Review/validate required technology to support service delivery for BIM compliant projects.
- Engage with your clients.
- Innovate.
- Embrace change!

The following, rather thoughtprovoking quotes perhaps best capture the underlying impact of BIM, and offer a fitting note on which to end:

> The world as we have created it is a process of our thinking. It cannot be changed without changing our thinking.
>
> You have to learn the rules of the game. And then you have to play better than anyone else.
>
> Albert Einstein

References

HM Government (2015) *Digital Built Britain: Level 3 Building Information Modelling – Strategic Plan*, BIS, London, UK.

Morrell. (2015) 'Collaboration for change: The Edge Commission report on the future of professionalism'. The Edge. Online. Available HTTP: <http://www.edgedebate.com/wp-content/uploads/2015/05/150415_collaborationforchange_book.pdf >.

NBS (2015) *NBS National BIM Report*. RIBA Enterprises, Newcastle, UK.

Walshe, G. and Daffern, P. (1990) *Managing Cost-Benefit Analysis*, MacMillan, Basingstoke, UK.

Glossary

Attributes: piece of data forming a partial description of an object or entity (PAS1192-2:2013).

Augmented/immersive technologies: technology that blurs the line between the physical world and digital or simulated world, thereby creating a sense of immersion (Wikipedia).

Augmented Reality: a composite view providing a combination of the real scene viewed by the user and a virtual scene generated by the computer that augments the scene with additional information.

Building Information Modelling (BIM): an holistic approach to the design, construction and management of facilities in the built environment. It comprises information, process and technology, brought together with the aid of three dimensional, real-time, dynamic building modelling software to create models, that encompass geometry, spatial relationships, geographic information, and quantities and properties of facility components using a shared information model environment (CDE). A collaborative way of working, underpinned by the digital technologies which enable more efficient methods of designing, delivering and maintaining physical built assets.

BIM authoring tool(s): any software tools used to create 3D BIM models.

Built Environment as a Service: the concept of providing an asset or part of the built environment as a service, with ownership and responsibility retained by the provider (or supplier), rather than being passed to the client (or user) on completion of the delivery contract. A model found in other industries, for example, in aircraft manufacture, where major components such as engines, are effectively leased by the airline, with the engine manufacturer retaining ownership and responsibility for operational maintenance, often for the life of the component

COBie drops: an exchange of information (data) at an agreed stage (milestone) of the project

Cost: an amount that has to be paid or spent to buy or obtain something. This might apply to the cost of part or all of an asset. Costs are typically incurred throughout the supply chain and incurred at the point of each transaction i.e. between sub-contractor and contractor, or contractor and the client. It is essentially the amount paid for goods or services rendered.

Cost assemblies: cost of a component or sub-component; an example might be the cost of a door or window component.

Disputes referee: person or persons appointed to review and advise on the resolution of construction contract disputes.

Employer: the term used universally to mean the facility and/or construction client.

(The) Engineering and Construction Contract: a standard form of construction contract published as part of the NEC 3 suite of contracts by the Institution of Civil Engineers, updated and re-issued in April 2013.

Federated 3D Model: an assembly of distinct models to create a single, coordinated and complete model of a building or asset. It enables collaborative working through separate, but coordinated models.

Intermediate consumption: value of the goods and services consumed as inputs by a process of production (OECD).

Level 2 BIM: the UK Government target for all centrally procured public sector projects requiring project participants to provide defined outputs via a BIM and that the resulting combined BIM will be managed as a series of self-contained (federated) models using proprietary information exchanges between systems.

Level 3 BIM: a single collaborative, cloud based, building information model.

Level of Detail: a term used to define the extent to which a model is developed as defined in the Protocol.

New Engineering Contract: a standard form of contract published by the Institution of Civil Engineers.

Objects: item having state, behaviour and unique identity – for example, a wall object (PAS1192-2:2013).

Object-oriented: typically used to define a type of computer programming language supporting the use of hierarchical entities and objects.

Optioneering: assessing alternative design options.

Parametric models/modelling: use of parameters (numbers or characteristics) to determine the behaviour of a graphical entity and to define relationships between model components, i.e. size of a window or door opening linked dynamically to the size of the window or door.

Price: the amount of money expected, required, or given in payment for something.

Professional Services Contract: a standard form of professional appointment published as part of the NEC 3 suite of contracts by the Institution of Civil Engineers updated and re-issued in April 2013.

Project Team Member: the term used within the Protocol to define a party with modelling obligations on a project.

The Agreement: the term used within the CIC BIM Protocol to define the underlying contract (such as the construction contract or professional appointment) in which the Protocol is incorporated.

The CIC Protocol: the CIC's Building Information Management Protocol first edition 2013 available free to download from www.cic.org.uk

Risk Analysis: the assessment of the severity of a risk.

Risk Management: the process of controlling the impact of risk.

Rule sets: a set of rules applied to control a given outcome, i.e. rules governing design.

Service Contract: term typically applied to services required, for example to support the operation and maintenance of an asset, i.e. cleaning services.

Spatial coordination: ensuring that design is coordinated to avoid clashes within any part of the resulting model.

Transactional contracts: traditional contract between two parties for the provision of works, goods or services for an agreed fee or price.

Value: assessment of the benefits brought by something in relation to the resources needed to achieve it.

Value Management: umbrella term used to embrace all activities and techniques used in the effort to deliver better value for the client.

Index